Intelligent Systems Reference Library

Volume 252

The aim of this series is to publish a Reference Library, including novel advances and developments in all aspects of Intelligent Systems in an easily accessible and well structured form. The series includes reference works, handbooks, compendia, textbooks, well-structured monographs, dictionaries, and encyclopedias. It contains well integrated knowledge and current information in the field of Intelligent Systems. The series covers the theory, applications, and design methods of Intelligent Systems. Virtually all disciplines such as engineering, computer science, avionics, business, e-commerce, environment, healthcare, physics and life science are included. The list of topics spans all the areas of modern intelligent systems such as: Ambient intelligence, Computational intelligence, Social intelligence, Computational neuroscience, Artificial life, Virtual society, Cognitive systems, DNA and immunity-based systems, e-Learning and teaching, Human-centred computing and Machine ethics, Intelligent control, Intelligent data analysis, Knowledge-based paradigms, Knowledge management, Intelligent agents, Intelligent decision making, Intelligent network security, Interactive entertainment, Learning paradigms, Recommender systems, Robotics and Mechatronics including human-machine teaming, Self-organizing and adaptive systems, Soft computing including Neural systems, Fuzzy systems, Evolutionary computing and the Fusion of these paradigms, Perception and Vision, Web intelligence and Multimedia.

Indexed by SCOPUS, DBLP, zbMATH, SCImago.

All books published in the series are submitted for consideration in Web of Science.

Christophe Gaie · Mayuri Mehta
Editors

Transforming Public Services—Combining Data and Algorithms to Fulfil Citizen's Expectations

Springer

Editors
Christophe Gaie
French Prime Minister Services
Paris, France

Mayuri Mehta
Department of Computer Engineering
Sarvajanik College of Engineering
and Technology, Sarvajanik University
Surat, India

ISSN 1868-4394 ISSN 1868-4408 (electronic)
Intelligent Systems Reference Library
ISBN 978-3-031-55577-0 ISBN 978-3-031-55575-6 (eBook)
https://doi.org/10.1007/978-3-031-55575-6

This Springer imprint is published by the registered company Springer Nature Switzerland AG
The registered company address is: Gewerbestrasse 11, 6330 Cham, Switzerland

Paper in this product is recyclable.

Preface

In recent decades, public organizations have earnestly embraced digitalization to foster e-Government, transforming traditional administrative procedures into accessible online forms. Consequently, most administrative processes are now explained on public websites and can be accessed online. This modernization is applied to existing procedures by converting paper to electronic form. Today, electronic procedures are available for declaring annual income, paying taxes or traffic fines, registering at school, high school or university, obtaining social subsidies, retrieving unemployment benefits, and so on.

The transition to e-Government has been accelerated with the development of Information and Communications Technologies, yielding manifold advantages. Firstly, digital procedures enhance public service efficiency and empower citizens to independently navigate processes with 24/7 accessibility. Secondly, modernizing public services improves citizen satisfaction, as they can complete procedures independently with 24-hour/7-day availability, eliminating the need to visit distant administrative offices. Thirdly, citizens gain more information about their rights. Forth, public services offer higher transparency through extensive online documentation. Finally, public policies become more effective, impacting a large part of the population at a lower cost.

However, e-Government is still not satisfactorily implemented, and some limitations persist. The classic approach of replacing the former paper-based procedure with a simple electronic form is a significant limitation. While electronic forms simplify the user experience and reduce civil servants' tasks, they do not modernize the process. For instance, tax and customs declarations still require companies to hire employees to fulfil their obligations or develop IT software. Thus, a new automated process is essential, that independently computes and collects tax fees. Additionally, modernizing procedures often coincide with reducing public service establishments.

Presently, the services provided to citizens are perceived as distant and less personalized. This poses new challenges in guiding and assisting people who suffer from digital illiteracy, affecting more than 30% of the population of developed countries and hindering the adoption of digital public services. Consequently, it is crucial to consider citizens' education and ensure personalized, exceptional procedures for disadvantaged populations. Furthermore, e-Government does not address the "non-take-up problem", where citizens may not fulfil administrative procedures to obtain a subsidy if the effort does not seem worthwhile, especially for low and non-recurring amounts such as specific checks for mobility or energy consumption.

To address the abovementioned limitations, the authors propose accelerating the transformation of public services and transitioning from Electronic Government to Automatic Government ("a-Government"). This new paradigm offers higher benefits, having the government service itself performs administrative procedures from the citizen's perspective. Simultaneously, it also alleviates the burden and ensures that every eligible citizen benefits from subsidies. Moreover, the automatic approach enhances the efficiency of public services, requiring better data exchange between public and private services, reducing fraud, and providing better support for low-income people.

This book delves into the a-Government approach, outlining strategies, secure architectures, and practical implementations through concrete use cases in education, agriculture, and human resources. It provides detailed methods and approaches to facilitate their implementation in other government organizations and inspires civil servants to develop new digital services for the benefit of citizens.

Paris, France
Surat, India
January 2024

Christophe Gaie
Mayuri Mehta

Contents

1 **Digital Transformation of Public Services: Introduction, Current Trends and Future Directions** 1
Christophe Gaie and Mayuri Mehta

2 **Enhancing Citizen Satisfaction Using Citizen-Facing Process Mining** 29
Shreekanth M. Prabhu, Natarajan Subramanyam, and V. Jagadishwari

3 **Thinking Inside the Sandbox: Beyond Public Services Digitalization with Co-Production** 87
Polyana Batista da Silva and Fabrício Ramos Neves

4 **Assessing Sustainable Development of e-Government Services Towards Citizen Satisfaction Through Continuous Improvement Approach** 107
D. Vimala, S. Vasantha, and A. Shanmathi

5 **Usage of Modern API for Automization of Government Procedures** 131
Rahul Arvindkumar Vaghela, Kamini Solanki, Rimpal R. Popat, Ila Rahul Vaghela, and Nidhi Chhangani

6 **Epidemiology Inspired Cybersecurity Threats Forecasting Models Applied to e-Government** 151
Jean Langlois-Berthelot, Christophe Gaie, and Jean-Fabrice Lebraty

7 **The Provision of e-Services by Public Administration Bodies and Their Cybersecurity** 175
Christophe Gaie and Mirosław Karpiuk

8 Modernizing the Processes for Human Resources Through Artificial Intelligence to Optimize a Variety of e-Government Responsibilities 189
Christophe Gaie and Markus Mueck

9 Fuzzy Logic Architecture for Availing E-Governance Healthcare Services by Rural Citizens 207
M. Bhuvana and A. Ramkumar

10 Enhancing the Efficiency and the Security of e-Government: The French Case Study of Human Resources Applications 223
Christophe Gaie

Chapter 1
Digital Transformation of Public Services: Introduction, Current Trends and Future Directions

Christophe Gaie and Mayuri Mehta

Abstract The digital transformation of public services is a key concern for government administrations as it optimizes services delivered to citizens, enhances the job of civil servants and increases the efficiency of government services. This chapter proposes an overview of the opportunities and challenges to consider for the implementation of e-Government. First, the concepts towards e-Government are defined to set common guidelines to the current book. Then, the chapter describes the challenges to implement e-Government that requires implementing an adequate ICT infrastructure, ensure security and privacy, acculturate the senior management and citizens to technology, and communicate on the availability of digitalized services. Afterwards, the authors describe the challenges of implementing cloud computing for government services notably in terms of security, privacy, technology readiness, and reliability. Finally, the authors detail new innovations that will radically transform the domain such as collaboration e-Government, Green IT, digital twin usage, or gamification. Throughout this chapter, the authors underline the importance of adopting a user-centric approach to fulfill the highest priorities of e-Government stakeholders.

Keywords e-Government · Public transformation · Digital transformation

C. Gaie (✉)
Chief of Engineering and Innovation Department, French Prime Minister Services, 75007 Paris, France
e-mail: christophe.gaie@gmail.com

M. Mehta
Computer Engineering Department, Sarvajanik College of Engineering and Technology, Surat 395001, India
e-mail: profmayurimehta@gmail.com

C. Gaie and M. Mehta (eds.), *Transforming Public Services—Combining Data and Algorithms to Fulfil Citizen's Expectations*, Intelligent Systems Reference Library 252, https://doi.org/10.1007/978-3-031-55575-6_1

1.1 Introduction

From many years, the interest towards digitalizing government services is continuously increasing. As a matter of fact, Information and Communication Technologies (ICTs) are constantly improving and offering new opportunities in terms of data exchange, reutilization, computation, combination, etc. In the meantime, there is also a growing demand towards obtaining new services to perform public procedures everywhere, anywhere and above all easily. Each person involved in this profound change wonders whether Digitalization transforms Public Services or the reverse. First, let's define what e-Government refers to.

According to the United Nations, electronic government refers to the "use of ICTs to more effectively and efficiently deliver government services to citizens and businesses" [1]. This means that e-Government plays a role in the transformation of public services by facilitating their usage and increasing citizen acceptation. While e-Government was initially a domain reserved to specialists, the scope is now very wide with many dimensions tackled.

The emergence and growth of e-Government is a long process. Indeed, it requires gathering public policy requirements, human IT skills, information and communication technologies, as well as financial means to modernize government services. If one of these requirements is missing, then e-Government cannot flourish. The following figure illustrates this approach (Fig. 1.1).

Implementing e-Government also relies on a progressive transformation [2–4] that can be modelled in five stages of increasing maturity, from the web appearance with low services and interactions, to the seamless web with rich services, modern technologies and multiple citizen interactions.

The first stages of e-Government improved the efficiency of public services in terms of productivity, swiftness and preciseness as paper-based procedures were digitalized [5, 6]. However, it did not totally transform the perception of public services by citizen that still find public services complex and time consuming [7]. This explains the importance of underlining the importance of transforming public

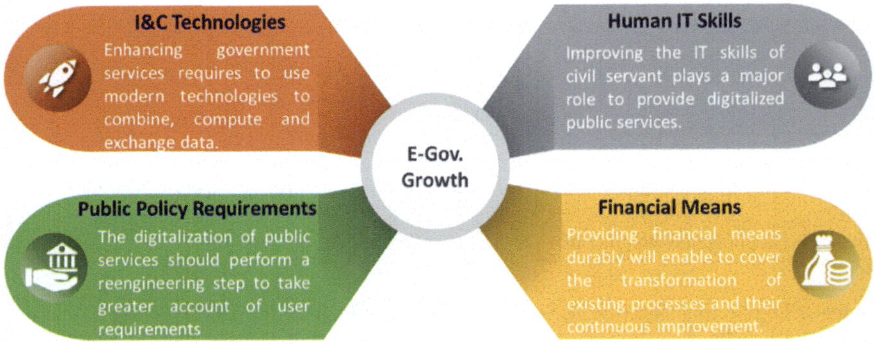

Fig. 1.1 The necessary ingredients to develop e-Government

services by a citizen centric approach such as the e-Tax approach in Japan [8] and measuring citizen satisfaction [9].

While digital transformation and e-Government are largely tackled in the literature, there are many confusions between related concepts such as e-Participation, eGovernance, etc. As a consequence, a first objective of this chapter is to define these concepts and facilitate their understanding. A complementary goal of this chapter is to detail the current trends of the domain, especially the relations with the major trends of Information and Communication Technologies such as artificial intelligence, cloud computing, or cybersecurity.

1.2 Evolution of the e-Government Concepts

In this section, the authors define multiple notions necessary to cover the domain of e-Government. First, we explain and illustrate how e-Government expanded to tackle new concerns and offer better perspectives (the figure is inspired from [10]) (Fig. 1.2).

There are numerous publications concerning e-Government that is the common concept do describe the usage of ICT to improve government procedures and replace classic paper-based public services. A description of the concepts of e-Government was proposed by Fang [11] and gives the definition of e-Government, the types of e-Government, the differences with e-governance, the interactions between government, business and the civil society, the Research Issues in Public Administration, etc.

A recent literature review was proposed by Liva et al. [12] and provides a landscape of recent research concerning e-Government. The paper outlines that there are many challenges to overcome above technology in order to transform government processes. As a matter of fact, it is required to take into account organizational,

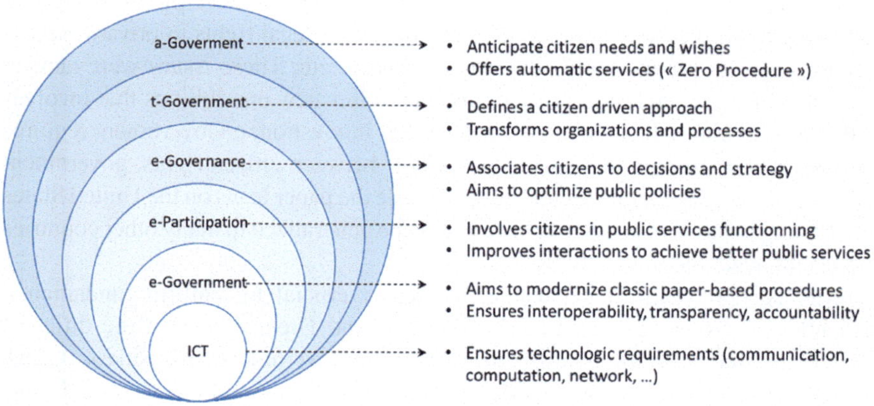

Fig. 1.2 The evolution of government approaches

juridical and social aspects to ensure that a new equilibrium is defined to achieve government procedures.

The following subsections hereafter describe in details each step concerning the evolution of government approaches to provide a strong theoretical background of the domain. In each subsection, the authors first describe the notions, then provide a bibliography of relevant papers in the literature, and finally provide a table that compares the content of cited papers.

1.2.1 About e-Government

Definition: e-Government refers to the initial approach of the domain that consists in modernizing public services or procedures thanks to the emergence of Information and Communication Technologies

Further information: e-Government is continuously enriched with new research directions (e-Participation, e-Governance, t-Government, etc.) Thus, e-Government may also refer to the comprehensive field of research

After several years of research and development of e-Government, Grönlund and Horan provided a relevant review concerning the domain that may be considered as a landmark [13]. The paper provides useful definitions and references to structure the different approaches adopted by researchers. A very important teaching relies in the understanding that e-Government is a cross-domain research field that notably combines legal aspects, public administration, computer science.

A key aspect in developing e-Government consists in guaranteeing to citizens that technology will preserve, and possibly improve, their legal rights to privacy, access to public services, reasonable government control, etc. These issues were tackled by Reitz [14], who underlines that technology raises new possibilities that involves both opportunities and challenges. As any other innovation, e-Government requires a transparent regulation that ensures a tradeoff between citizen rights, government efficiency as well as private sector growth. While the paper leans on the United States functioning in the 2000s, the conclusions may be generalized to many other countries and is still available in the 2020s.

After several years of expansion, it became crucial to reaffirm fundamental principles of e-Government. In [15], Alshehri and Drew described the different services offered: Government-to-Citizen (G2C), Government-to-Business (G2B), Government-to-Government (G2G), and Government-to-Employee (G2E). It also describes the benefits of e-Government as well as the multiple barriers to overcome:

Table 1.1 Summary of the previous work concerning e-Government

e-Government		
Id	Contribution	Limits
[13]	Provides useful definitions and references to structure e-Government	The number of references is limited to be considered as a review
[14]	Outlines the importance of a regulation concerning data used for e-Government	There is no concrete proposal to protect data
[15]	Details the principles of e-Government and identifies barriers	There is no proposal to overcome the difficulties encountered to implement e-Government
[16]	Describes the factors that enable the adoption of e-Government	The research does not tackle how to benefit from the identified enablers
[17]	Outlines that e-Government is facilitated by precise factors	A model is detailed but not concretely applied

infrastructure, privacy, security, regulation, training, partnership, digital divide, culture, management support. This paper plays a major role to understand the foundations of e-Government.

The experience towards e-Government also offers the opportunity to analyze and identify the factors that facilitates its adoption by citizens. As a matter of fact, Gupta et al. underlined that the adoption of e-Government is facilitated by performance and effort expectancy, social influence, trust in government and technology as well as citizen satisfaction [16]. These findings were obtained thanks to the implementation of a model that relies on the unified theory of acceptance and use of technology (UTAUT).

A comprehensive approach of e-Government was proposed by Malodia et al. that describe the evolution of e-Government in five successive stages, provide the results of interviews from various stakeholders (government officials, intermediaries, citizens and ICT experts), outline that the success of e-Government is facilitated by citizen orientation, channel orientation and technology orientation and moderated by the digital divide, economic growth and political stability [17]. The authors propose a model that rely on the Gioia method [18] and detail the outcomes of e-Government but do not apply it concretely. This is a perspective of research of their work.

The summary of the proposed literature concerning e-Government is provided in Table 1.1.

1.2.2 About e-Participation

Definition: e-Participation is a branch of e-Government that aims to involve citizens in the design and delivery of government-related processes

Further information: developing e-Participation implies to share the construction and the usage of public services with citizens. Involving citizens is a powerful mean to accelerate the modernization of government services

In the literature, the first experimentations to facilitate the emergence of e-Participation consisted in proposing a platform that enable the collaboration between civil servants and citizens [19]. Anthopoulos et al. describe how to implement these platforms operationally and detail the possible connection to future e-Government systems. This approach is intended to enhance the participation of citizens to the design of public services and improve their adoption.

A useful comparison between e-Government and e-Participation was proposed by Peristeras et al. [20]. The paper outlines that they both combine Society and Governance with the clarification that e-Government is associated to the government system part of Governance while e-Participation is linked to the political system of Governance.

The maturity of e-Government and e-Participation was examined by Krishnan et al. using a multiple mediation model [21]. The proposed model defines e-Participation as a combination of e-Information Sharing, a-Consultation and e-Decision Making. The authors applied their model to a set of data from 187 countries and outlined that technological infrastructure, human resources and e-participation are closely correlated to e-Government maturity.

The interest towards e-Participation was detailed [10] and the authors identified five core capabilities that enable citizens to achieve create value through e-Participation: ICT access, social interaction opportunities, equality between citizens, freedom of expression, and capability to acquire knowledge. The research outlines the current limits of e-Participation and recommend strengthening the core set of capabilities identified to make a further step forward.

After several years of expansion, a useful review of e-Participation was proposed Steinbach et al. [22]. The authors proposed a detailed approach that structures the domain in three successive phases: the adoption of e-Participation, its implementation and its institutionalization that aims to maintain it on the long-term. They also classify research depending on the level of analysis from the thinnest dimension of individuals to the systemic viewpoint of the organization; most of the literature is dedicated to the adoption and implementation of e-Participation at the macroscopic level of administration and at the organizational level.

Another valuable state-of-the-art concerning e-Participation was made by LeBlanc in [23]. It outlines the creation of multiple platforms dedicated to this objective and the inefficiency of this strategy. As a matter of fact, there is a need to better take into account citizens' expectations as well as the reluctance of government services to share with citizens the initiative of policy making. The author also proposes some ideas to associate citizens to policy-making processes.

The summary of the proposed literature concerning e-Participation is provided in Table 1.2.

Table 1.2 Summary of the previous work concerning e-Participation

e-Participation		
Id	Contribution	Limits
[19]	Proposes an implementation of e-Participation based on e-Government Groupware	The acceptance of inter-service collaboration is relatively low. This reduces the capability to deploy widely the system proposed
[20]	Provides a comparison between e-Government and e-Participation	The paper does not provide operational material to develop e-Government or e-Participation
[21]	Describes a model to define the interactions between source factors, e-Participation and e-Government	The approach is mathematical and not directly operational in a government context
[10]	Identifies core capabilities to facilitate the development of e-Participation	The proposal is not directly applicable in another context
[22]	Provides a classification of papers concerning e-Participation	The state of the art does not fill the gap identified
[23]	Underlines the importance to better take into account citizens' expectations and provides some proposals to improve it	There is no proposal

1.2.3 About e-Governance

Definition: e-Governance extends the domain of e-Government to the governance of ICT, to the satisfaction of citizens and their association to the decision process

Further information: e-Governance is a two-way protocol while e-Government is a one-way approach based on the provision of public services

The definition of e-Governance and its main contribution were detailed in [24]. As a matter of fact, Heeks underlined the improvement of government processes (e-administration), the better connection of citizens (e-services), and the emergence of external interactions (e-society). The author details many case studies to outline the limits of e-Governance. He also identifies the initiative to follow in order to improve e-Governance. While these orientations are mentioned, they are not specified in details.

The evolution towards e-Governance was also conceptualized by Finger and Pécoud in [25]. As a matter of fact, they outline the positive effects of new ICTs with 1–the mere substitution of procedures, 2–the mirroring approach to offer information to citizens, 3–the emergence of new services for citizens and 4–the better efficiency of public services delivery. The authors propose a model where e-Governance depends on four cumulative factors (evolution of ICT, geographical level, functions to fulfill, and involved actors). While this model offers a valuable modelization of e-Governance, the research does not offer concrete implementation.

Rossel and Finger proposed another conceptual approach where e-Governance is a tradeoff between technology and institutional transformation [26]. They analyzed e-Governance through the prism of six dimensions, three types of actions, three geographic levels, and three functions that are similar to the approach of [25]. While this model is relevant to determine the factors involved in e-Governance, the research does not apply them to a concrete case study.

The difference between e-Government and e-Governance was also descripted by Hassan who underlined the fact that governance refers to a structure where institutions, organizations and citizens define a way to interact without being directed by the government [27]. The author also points out that e-Governance offers better services, information and transparency to citizens. This contributes significantly to democracy. While e-Government is illustrated by the case study of Bangladesh, there is no example for e-Governance.

The history of the evolution from e-Government to e-Governance was also explained by Dawes who underlines four major issues in [28]: 1–executive priorities[1]; 2–legislative concerns[2]; 3–technology leadership[3] and 4–Government-wide concerns.[4] By tackling these issues, e-Governance brings substantial improvements such as the emergence of a policy framework, the development of better public services, the better efficiency of services by considering users' requirements, and finally, the improvement of citizen engagement. While the research is very useful to understand the evolution of the domain, it does not provide new proposals to develop e-Governance.

Finally, a very innovative approach that consists in proposing a platform that enable to implement concretely e-governance was proposed by Lee-Geiller and Lee in [29]. As a matter of fact, the authors propose the Democratic E-governance Website Evaluation Model (DEWEM) that relies on measure indicators such as Ease of Use, Capability of Interaction, Portal convenience, Privacy policy, Trust in portal, etc. Interestingly, an implementation of the DEWEM architecture is proposed and details the different steps to fulfill. Thus, the DEWEM was implemented with cumulative steps (preliminary studies, preparation of data, credibility check and final improvement. While the model is fully valuable, it was not tested in operational situations.

The summary of the proposed literature concerning e-Governance is provided in Table 1.3.

---

[1] Examples: education, security, economy.

[2] Examples: identity, privacy protection, data regulation.

[3] Examples: Green IT, IT security, Heath IT.

[4] Examples: digital divide, privacy, governance.

Table 1.3 Summary of the previous work concerning e-Governance

e-Governance		
Id	Contribution	Limits
[24]	Details many case studies to outline the limits of e-Governance and proposes improvements	Does not detail improvements operationally
[25]	Outlines the positive effects of new ICTs for e-Governance and proposes a model	Does not implement concretely the proposal
[26]	Analyses e-Governance through different dimensions and identify the contributing factors	Does not apply the factors to concrete e-Governance use cases
[27]	Underlines the interactions between institutions, organizations and citizens in e-Governance	Does not provide examples concerning e-Governance
[28]	Identifies four major issues explaining the evolution towards e-Governance and identifies the different improvements	Does not provide new ideas to improve e-Governance
[29]	Proposes a model to evaluate e-Governance (DEWEM) and explain how to implement it	Does not apply the model to concrete case studies

1.2.4 About t-Government

Definition: t-Government refers to the shift from a digitalization of government services as performed by e-Government, to an in-depth transformation of the processes to fulfill citizen requirements and offer them the opportunity to be an actor of public policy making

Further information: t-Government goes further than e-Governance in the association of citizens as it offers them the opportunity to modify public procedures and propose new simplified approaches

The emergence of t-Government is a recent notion that refines the classic approach of e-Government. Irani et al. underline three directions carried out by t-Government: 1—focus on citizens needs and requirements to deliver public services; 2—develop a common culture of service[5] to optimize the final value offered to citizens; and 3—improve the management of resources and skills to develop professionalism of public service delivery [30]. The paper offers crucial definitions, model and directions to initiate a dynamic towards the emergence of t-Government. However, there is a need to apply it to concrete case studies to verify the effectiveness of proposals.

[5] This could be expressed as "Satisfy a client rather than deliver a service".

Switching from e-Government to t-Government is not an easy process as it requires to better take into account the needs and expectations of citizens. Tony Blair, the former Prime Minister of the United Kingdom, gave an impulse to promote t-Government as indicated by King in [31]. The research also highlights the importance of developing a significant citizen-focused technology such as the customer relationship management (CRM) and proposes a four-stage maturity model. Interestingly, the author underlines a difficulty between stages three and four, as there is a barrier that is related to the traditional, bureaucratic government organizations. Unfortunately, the author does not indicate how to overcome this difficulty.

Above the idea of transforming government services through a user-centric perspective, t-Government also aims to improve transparency, accountability, flexibility and cost efficiency of public services. To fulfill these objectives, Janssen and Shu propose in [32] a framework based on three major axes: 1—defining the types of transformations to perform, 2—addressing the levels of abstraction involved and 3—choosing powerful supporting instruments. Although the framework is very helpful to implement t-Government, it is not immediately applicable as there is a need to define a methodology and to experiment it on a concrete use case.

After two decades of research concerning t-Government, Omar proposed a large review of the domain and highlighted the challenges, complexities and differences of approach in [33]. The authors analyzed the word frequency of 496 research papers in the domain and pointed the 50 most frequent words (the first one is "government", then "public", "information" …) and built a word tree of research topics. While this paper provides a very important review of the domain, it does not propose a clear classification that can be reused in further research.

The new concept of digital government ambidexterity (DGA) was imagined by Björses and Ivarsson in [34] and refers to capacity to take simultaneously advantage of incremental and radical digital innovations to bring value to public service. This concept was modelled and optimized by Ndlovu et al. in [35] to verify the hypothesis that "Digital Government Ambidexterity influences Public Value Creation" using a sample of 690 citizens using digital government platforms. The research proved that DGA has a significant influence on information usability, completeness, readability, trustworthiness, and relevance. This means that DGA can play a major role in the development of t-Government to fulfill citizens expectations. While this research provides meaningful information, the methodology relies on the perfectible approach of self-completion questionnaires.

The summary of the proposed literature concerning t-Government is provided in Table 1.4.

Table 1.4 Summary of the previous work concerning t-Government

t-Government		
Id	Contribution	Limits
[30]	Defines t-Government and provides a model and directions to initiate a dynamic	Does not apply the proposals to concrete case studies to verify their effectiveness
[31]	Highlights the importance of developing a significant citizen-focused technology such as the customer relationship management (CRM) Proposes a four-stage maturity model	Does not propose a solution to the barrier that appear between stages three and four of the maturity model
[32]	Proposes a framework based on three axes	Does not implement t-Government on a concrete use case
[33]	Provides a very important review of the domain	Does not propose a clear classification that can be reused in further research
[35]	Offers meaningful information	The methodology relies on the perfectible approach of self-completion questionnaires

1.2.5 About a-Government

In this book we propose to introduce a new approach entitled "a-Government" (Automatic Government) that aims to provide automatically public services to citizens. The perspective pursued is to reduce the administrative burden supported by citizens as well as reduce the no recourse to public subsidies such as minimum revenue, tax credit, environmental aid, etc.

As a matter of fact, the modernization of government services enables to achieve many procedures by internet such as tax reporting, public institution enrolment, booking an appointment with a civil servant, obtaining a public subsidy, etc. This evolution clearly improves the interaction between citizens and government services by avoiding the former to go to the appropriate service and the latter to replicate data from a paper form into the appropriate database. However, citizens still have to achieve the procedure and some of them can also encounter electronic barriers due to digital illiteracy. This is the reason why a-Government will provide better user experience with, for example, reporting revenues automatically based on pay declaration of employers or granting subsidies based on the data already held by government services.

Another perspective towards a-Government is the capability to propose personalized services to citizens by taking advantage of Artificial Intelligence. Indeed, a valuable improvement would be to perform services from a simple event. For instance, a pregnancy declaration could imply the completion of a demand to obtain a place at the nursery; a birth could trigger the automatic creation of a social insurance record; reaching the age of the majority could result in a voter registration. The usage of AI could also enable suggestions such as proposing student accommodation to young people who register to the university, propose a medical appointment with a specialist when multiple medical acts are detected, etc.

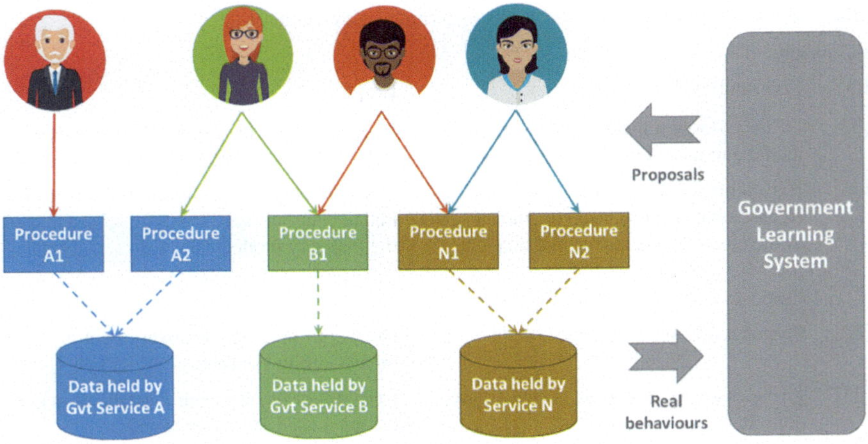

Fig. 1.3 A possible implementation of a-Government

An interesting approach could be to inspire from other citizens behaviors such as market places do. This refers to the principle of a similar track (i.e. "Other customers also bought these items") that can be learnt by a Government Learning System (Fig. 1.3).

1.3 The Public Value of Transforming Public Services Through e-Government

In this section, we describe the public value of e-Government based on the existing literature. A key point consists in identifying the levers that encourage to modernize public services and favor e-Government. Karunasena et al. proposed a very useful framework that identifies four sources of public value creation from their literature review [36].

1. **Delivery of public services**: this perspective evaluates notably the quality of information delivered to citizens, the amount of data available, and the simplicity of public service usage. It aims to measure the satisfaction of citizens towards public services.
2. **Achievement of outcomes**: this dimension aims to measure the social outcomes obtained from the modernization of public services. The evaluation takes into account the impacts, the deliverables, and the consequences of e-Government for the society.

The authors of this chapter consider that other outcomes should also be considered. For instance, the impacts on the environment (i.e. greenIT) should be considered.

3. **Development of trust**: this perspective enables to examine whether e-Government offers an adequate protection of personal information (security and privacy), details the process of public decision making (transparency), inspires confidence in public services (trust), and contributes to enhance the involvement of citizens in governmental action.
4. **Effectiveness of public organization**: this dimension aims to ensure that e-Government improves the functioning of public services through cost reduction, better performance, and higher inter-service collaboration. For each significant evolution, a measure of the financial return of investment in public organizations is required to assert the effectiveness of the organization.

An interesting method to combine and compute the value of these levers was proposed by Mensah et al. [37]. In their research, the authors propose a public value of e-Government adoption model (PVEGAM) that combines the data using seven hypotheses whose validity was asserted using a questionnaire completed by Chinese. This survey contributed to evaluate the relation between levers and public value of government (mathematical correlation). The highest contribution comes from the efficiency and just behind it the quality of information.

Finally, a crucial approach concerning the public value of e-Government consists in evaluating the perception from different point of views. Indeed, as described by Castelnovo, the citizen is alternatively a decision-maker, a civil servant, a participant of a process, or a taxpayer [38]. As a consequence, the public value perceived by a given stakeholder depends on the role it endorses.

To initiate new research approaches, we propose here to model the problem as an optimization problem. The objective pursued is to maximize the return on invest (ROI) of an effort to increase public value (i.e. the sum of each factors described by Karunasena, Mensah, or any other researcher) while ensuring a minimum value for each stakeholder. Define the following variables.

- i a given improvement of the public process
- $d(i)$ the delivery gain obtained from improvement i
- $o(i)$ the outcome gain obtained from improvement i
- $t(i)$ the trust gain obtained from improvement i
- $e(i)$ the effectiveness gain obtained from improvement i
- $c(i)$ the cost of improvement i.

Thus, the ROI of improvement i may be expressed as:

$$ROI(i) = d(i) + o(i) + t(i) + e(i) - c(i)$$

Consequently, a first objective would be to identify the list of improvements {i} that maximizes the ROI:

$$\underset{\{i\}}{\text{argmax}} \sum_{\{i\}} [d(i) + o(i) + t(i) + e(i) - c(i)]$$

This objective may be easily refined by introducing the stakeholder perspective. To this end, we define:

- $[AD] = [ADd, ADo, ADt, ADe]$ the vector that defines the level of acceptance on each factor for a decision maker
- $[AP] = [APd, APo, APt, APe]$ the vector that defines the level of acceptance on each factor for a participant
- $[AT] = [ATd, ATo, ATt, ATe]$ the vector that defines the level of acceptance on each factor for a taxpayer
- $[AS] = [ASd, ASo, ASt, ASe]$ the vector that defines the level of acceptance on each factor for a civil servant

As a consequence, the problem to solve is:

$$\underset{\{i\}}{\operatorname{argmax}} \sum_{\{i\}} [d(i) + o(i) + t(i) + e(i) - c(i)]$$

With the following constraints:

- $d + \sum_{\{i\}} d(i) \geq \max(ADd, APd, ATd, ASd)$
- $o + \sum_{\{i\}} d(i) \geq \max(ADo, APo, ATo, ASo)$
- $t + \sum_{\{i\}} d(i) \geq \max(ADt, APt, ATt, ASt)$
- $e + \sum_{\{i\}} d(i) \geq \max(ADe, APe, ATe, ASe)$

Finding the optimal combination of improvements {i} to maximize the ROI is a classic optimization process. The method to solve this problem relies on convex optimization and is detailed in [39], multiple examples are provided in [40] and [41].

A key point of this approach is to define suitable constraint vectors to ensure that the problem may be solved. To verify the convergence of the algorithm, setting the constraints at a low level is recommended.

1.4 e-Government Current Challenges

A key concern of e-Government is the capability of government services to move from theory to practice. As a matter of fact, the context of public services is currently very constrained in many dimensions. The succesive financial crises reduced the financial means and human resources of ICT government services while requiring higher results in terms of efficiency, data protection, intelligibility, interactivity, and service coverage.

These challenges applies to every country whether it is supposed to be rich, in development or poor since the barriers to implement e-Government apply in any situation. To illustrate the current challenges for e-Govnerment we propose the following figure inspired by Meiyanti et al. in [42] (Fig. 1.4).

While the next section is dedicated to highlight the current trends to enhance ICT government services, it is crucial to underline the importance of taking into

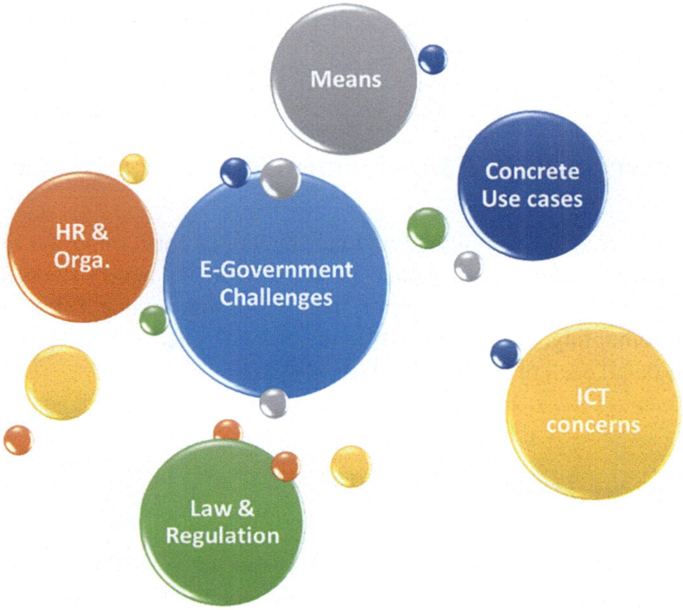

Fig. 1.4 Current challenges to develop e-Government

account concrete use cases. As a matter of fact, IT engineers and researchers often tend to solve theoritical problems that stillhave to be adjusted to real-life situations. This explains why we pay particular attention to detail the implementation of e-Government in concrete use cases such as education [43, 44], healthcare organization [45, 46], agriculture efficiency [47], or fraud detection [48].

A simpler model that identifies barriers and challenges to implement e-Government was proposed by Arief et al. in [49] based on a limited sample of 20 articles. The model relies on three major domains: technology, organization and human resources. The research outlines 34 concerns where the most recurring barriers are the lack of IT infrastructure, the insuficcient IT security, the excessive financial constraints, and the policy. While the analysis is promising, it should be further detailed by including other valuable papers, providing a deeper analysis of barriers and proposing levers to overcome them.

A valuable analysis of current trends in e-Government was proposed by Heidlund and Sunberg based on the keywords of 720 articles that were analysed using VOSviewer [50]. The research splits data into seven clusters: 1–general information concerning e-Government, 2–general information concerning e-Governance, 3–service quality, 4–calculus, 5–digital divide, 6–risk management, and 7–technology acceptance. The authors also analyze the content of the highest cited papers and conclude that a majority of papers focus on the quality of their service without analyzing government service as a whole. Thus, the author propose to extand classic approach so that researchers take better into account government policies from a systemic point of view.

1.5 e-Government ICT Development Trends at a Glance

In the previous chapter we outlined various challenges concerning the implementation of e-Government. The technical implementation is a very detailed axis of research of e-Government. It is currently focused on four major ICT trends: security, cloud computing, interoperability and artificial intelligence. The following subsections detail each of these technological trends.

1.5.1 Security Trends

In the current context of massive cyber-attacks, securing IT applications of government services represents a major priority. As underline by Efe and Kadzal, protecting IT applications of government services aims to ensure confidentiality, integrity and availability of sensitive information [51]. In this paper, the authors underline the importance to protect IT systems from external attacks that can reveal multifaceted (deny of service, virus, ransomware, ...). Classic protection measures should be implemented to protect government services such as firewalls, intrusion detection and prevention software, encryption and secure. Moreover, civil servants should be trained to cybersecurity as they can be a target of attacks or be actors of the cyber-protection.

Improving the security of e-Government requires following a global approach that begins by identifying the aspects to consider, goes on by proposing measures and finally analyzing the challenge to overcome. In [52], Yang et al. indicate that the dimensions involved to secure e-Government systems are technical, organizational, scientific, legal, economic and informational. Furthermore, the authors propose different measures to enhance the protection of IT government services such as blockchain to reinforce decentralized services, AI to detect spam, phishing or intrusions attacks, biometry to prevent the system from identity theft, etc. Finally, they detail the different challenge to overcome such as protecting wireless communication access, reducing the surface of attack, enhance spectrum efficiency to reduce the environmental footprint, or limit the application risks.

There is also a strong trend towards implementing blockchain in e-Government. For example, Khayyat et al. analyze the potential intakes of this technology provide the secure, transparent, and decentralized transactions [53]. Implementing blockchain for e-Government services would increase the reliability of data held (date of birth, marriage, death, address, identity, ...) that may be modified or corrupted at different stages of their lifecycle with a difficulty to detect the data corruption. The authors also outline the difficulties to overcome in the implementation process such as the lack of expert killed in government services, the insufficient regulation of the domain, and the lack of interoperability between blockchain technologies.

It is recommended to adjust the implementation of blockchain technology to the context of public services as it handles sensitive information about citizens, companies and other entities, and therefore is regularly attacked by cybercriminals. Thus, proposing a semi-decentralized approach such as suggested by Elisa et al. offers valuable advantages to secure the blockchain implementation [52]. The rights to perform transactions and/or manage the system are asymmetric and some selected actors are given the rights to create, submit, access, and review transactions.

A comparison of the blockchain solutions implemented in the context of e-Government was proposed by Kassen in [54]. The approach relies on the analysis of four case studies from different countries (Estonia, Germany and United Arab Emirates) and different areas (healthcare, asylum, smart city and military procurement). The author underlines the benefits from blockchain that improves public transparency by opening databases, accelerates decision-making processes and offers secured transactions.

To conclude this subsection, we outline valuable implementations of the blockchain in e-Government context.

- The attribution of frequency by the national frequency agency of France (ANFR) [55]
- Multiple practical applications were performed to fulfill the United Nations Sustainable Development Goals such as providing food, healthcare, or education [56]
- The production of property certificates for notarial office for residents in China [57].

These examples highlight the paradigm shift as blockchain becomes the reality for an increasing share of e-Government services.

1.5.2 Cloud Computing Trend

The usage of cloud computing is also a powerful trend of e-Government. Indeed, cloud computing offers significant benefits as it enables to deploy IT services easily, facilitates their usage by different stakeholders (internal, external, tier-party), and improves the scalability of services that can adjust rapidly to fluctuating requirements (peak activity, progressive growth, system redundancy, ...).

The challenges and benefits of implementing cloud computing for e-Government were analyzed by Al Mudawi et al. in [58]. Firstly, the paper describes the challenges to implement e-Government that requires to implement an adequate ICT infrastructure, ensure security and privacy, acculturate the senior management and citizens to technology and communicate on the availability of digitalized services. Then, the researchers describe the challenges of implementing cloud computing for e-Government services notably in terms of security, privacy, technology readiness, and reliability. Finally, the authors detail the contribution of cloud computing that we mentioned in the previous paragraph.

The connection between cloud computing and e-Government was also performed by Suleiman et al. who described the interest of combining these approaches [59]. As a matter of fact, implementing government services on a cloud infrastructure provides notably better elasticity of service, higher traceability, improved financial efficiency, and stronger resilience. However, they also detail the challenges to overcome to benefit from these new capabilities. To illustrate it by a metaphor "Being able to steer a boat on a lake does not guarantee that you can do it on the see" (the lake would correspond to an IT infrastructure belonging to government services while the see would refer to an external cloud service). This mean they are two main orientations to tackle: 1—taking advantage of the new capabilities offered by the cloud infrastructure, and 2—protecting the government IT systems from the new threats.

Cloud computing highly improves the sustainability of governance as it offers scalable and interoperable possibilities underlined by Fathey et al. [60]. Indeed, it provides better agility in the development process by enabling continuous delivery on multiple platforms that favors innovation. The authors provide a state-of-the-art that advocates for the implementation of cloud computing to improves the performance, reliability and scalability of e-Government services. While this paper offers an interesting overview of the domain, it does not bring new proposals to our point of view. Further research will probably extend the authors' useful analysis.

The emergence of cloudlets to facilitate the communication between mobile devices and clouds offers new opportunities to develop e-Government. Thus, Al-Shargabi et al. proposed a cloudlet model that ensures secured and trusted manner communications for government services [61]. The implementation relies on the combination of Simple Object Access Protocol (SOAP) and an encryption protocol (triple Data Encryption. Standard or Rivest-Shamir-Adleman standard). While the temporal performance is valuable, the results concerning trust are unfortunately not detailed. In any case, the trend towards implementing cloudlet for e-Government applications has to be considered in the upcoming years.

1.5.3 Interoperability Trend

Another important trend to favor the growth of e-Government is the capability to exchange data easily to combine data and provide advanced services to citizens. To reach this objective, government organizations pay attention to develop interoperable services as indicated by the review paper of Akbar et al. [62].

Estonia is very involved in the implementation of modern and digitalized government services. The functioning of the whole systems requires achieving interoperable services. Thus, the country adopted the X-Road path that requires fulfilling multiple prerequisites as described by Saputro et al. [63]: 1—a qualified signature with a strong authentication, 2—a unique administrative identifier, 3—a pre-existing information system, 4—an ICT security policy, 5—a digital records system, 6—a high-level sponsorship, and 7—a strong governing authority. The authors also underline the importance towards measuring the adoption of interoperable services through indicators of deployment. This offers the opportunity to publicize the project and its success.

The impact of interoperability in the adoption of digitalized public services in the European Union (EU) was analyzed by Campman et al. [64]. The authors perform an analysis based on empirical results that conclude in the positive effect for citizens of the partial adoption of the European Interoperability Framework. They propose some recommendations to pursue the interoperability progress such as strengthening the governance and introducing specific requirements for interoperability to be implemented in the EU to facilitate the measure of results. However, the authors also indicate that there is a need to ensure some flexibility for member states to take into account national specificities.

To increase the interoperability of public services, public organizations are currently implementing Application Programming Interfaces (APIs). They provide data that can be accessed using simple HTTP[6] requests that fulfill the requirements defined in the technical documentation (i.e. the swagger also entitled the OpenAPI Specification). Thus, Puspitasari et al. developed microservice APIs that can be used to interoperate in relaying data between e-Government services [65]. Microservice APIs also offer the opportunity to divide the development effort between multiple IT teams and enable the usage of various technologies, especially different coding languages.

Another approach consists in enabling the possibility to connect modern IT applications to the oldest ones (legacy systems) that can reveal difficult to rewrite. For example, the author of the current chapter described in [66] a method to connect the historic code of the French Tax Administration written in COBOL to new digitalized services that can use interactive technology such as Java, Php or React.JS. The research highlights the benefits of developing API over legacy systems so that new interactive systems can obtain data managed by the back-office systems. The author also describes a method to achieve a migration of the legacy that relies on the usage of a hybrid architecture.

Finally, a comprehensive description of the usage of API to modernize e-Government was proposed by Vaccari et al. [67]. The objective of the research is to provide a detailed documentation to facilitate the usage of API in every government services. To this end, the authors describe four complementary aspects: 1—the instruments offered by the European Commission to facilitate the emergence of APIs in public services, 2—the current technical standards of API, 3—an analysis of the existing API developed by European government services, and 4—a list of best practices for implementing API in government services. The results indicate that the topic is promising with many services to modernize through the web APIs standards that are currently used by government services.

[6] HTTP: HyperText Transfer Protocol.

1.5.4 Artificial Intelligence Trend

Artificial Intelligence is also an increasing trend of research in the domain of e-Government as it provides better public services in terms of efficiency, data valorization, citizen involvement, etc.

Implementing Artificial Intelligence to enhance e-Government is not a trivial task since the existing information systems are heterogeneous, not necessarily interconnected and do not rely on the same definitions of concepts. Thus, Berrada et al. propose a new architecture based on common ontologies that enable to exchange on a common basis [68].

The architecture described in their publication is composed of three layers: 1—the legacy layer, 2—the ontology layer, 3—the communication layer, and 4—the composition layer. A recommendation layer that would rely on machine learning based personalization is also detailed but not implemented. This proposal is promising and should be further investigated.

The use of AI can also facilitate how to manage information detained by government services [69]. As a matter of fact, Al-Mushayt proposed a framework to take advantage of the huge amount detained by government organizations. While the framework is not concretely implemented, the author also proposes a concrete set of deep learning techniques to automatize e-Government services. These AI techniques are applied to recognize hand-written information as well as classify opinions of citizens. Finally, the author details a new platform that enables the integration of new AI techniques in the e-Government systems and services to improve the efficiency of government procedures.

Furthermore, a very useful literature review was proposed by Zuiderwijk et al. and details the issues of AI in Public Governance [70]. The research classifies the research contributions according to the benefits they provide in terms of efficiency and performance, risk identification and monitoring, economic impact, data and information, quality of service, contribution to society, decision-making, citizen engagement, or environmental aspects. The paper also sorts papers according to the challenge they provide such as data management, organizational concerns, skill development, interpretability, ethics, legal or politic decision, societal dimension, and economic impact. This paper offers a larger overview of the opportunities and challenges to consider when introducing AI in the e-Government domain.

As observed in the previous paragraph, AI is a promising approach for e-Government. Thus, government services are looking to implement it rapidly and improve the government services. Yerlikaya and Erzurumlu describe the objectives that drive countries towards this objective that requires developing AI solutions in various contexts [71]: education, healthcare, cybersecurity, finance, etc. The paper highlights that AI should be applied to the public policies with the highest priorities for the population. The authors also detail the pillars that drive the application of AI for e-Government: 1—enhance the capability to acquire talent, enhance infrastructure, facilitate investment and improv governance, 2—build an ecosystem that favor the emergence of AI applications for the public and private sectors, 3—choose the use cases with the highest impact on society.

The potential applications of AI are numerous and were largely applied during the Covid-19 pandemic. For instance, Gaie and Mueck proposed an artificial intelligence framework that provides a tradeoff between sanitary and economic perspectives during the Covid-19 pandemic [72]. The research takes advantage of deep learning for medical applications and may be applied in other e-Government contexts. There are also many other potential uses of AI that can deeply change public and private activities such as outlined by Barthelmess in [73].

1.5.5 Other Innovative Trends

This last subsection describes new trends of research that could bolster the development of e-Government in the future.

To accelerate the transformation of public services, it is of highest importance to define a strategy and prioritize the tasks according to their value and cost of implementation. This observation advocates for considering with attention the research that aim to classify the actions to perform such as using as SWOT analysis as proposed by Karyono and Augustina in [74]. This approach offers concrete perspectives to transform government services to offer better public digitalized services to citizens.

Another interesting trend is the development of collaborative e-Government processes. This refers to mutually synchronized actions, by citizens, companies, associations, or other organizations to provide an input or output for e-Government. Gacitúa et al. provide a comprehensive review of the domain that tackle many complementary aspects involved in the process (adapted architecture, common concepts, normalized ontologies, shared models, etc.) [75]. While the interest towards collaboration is growing, it is also associated with others such as interoperability and required technologies to implement it.

To facilitate the adoption of e-Government, some researchers also examine the potential effects of gamification. For instance, Halsbenning et al. describe useful design principles to implement simulation game platforms (SGP) and enable citizens to experiment public services in a funny manner [76]. The authors provided a configurable SGP whose functioning was described in an Entity Relationship Model that describes the roles, events, actions that will occur in the game. This new approach of e-Government is promising and will undoubtedly contribute to facilitate the mutual understanding of producer (civil servants) and consumers (citizens, companies, other stakeholders) of public services.

While government services tend to facilitate the relationship between citizens and the government services, they usually are not built to anticipate citizens' expectations. As a matter of fact, it is difficult to anticipate their behavior since data held by government services are scattered in multiple IT systems. To waive this limit, researchers designed a digital twin that represents the real citizen by aggregating data that correspond to significant interactions with the government services. Anshari and Hamdan detailed the current innovation taking advantage of the digital twin to improve e-Government [77].

Lastly, the global warming catches the attention of policy makers towards reducing the carbon footprint of government services. However, Frogeri et al. highlighted that there is still low literature that tackles simultaneously e-Government and Green IT [78]. This failure will probably be filled in rapidly as the concern towards global warming is increasing and there is clearly room to optimize the efficiency of digitalized services efficiency. For example, there is a huge potential gain by simplifying administrative processes, optimizing administrative websites, improving data storage, or developing automatic procedures.

1.6 Conclusion and Perspectives

In this chapter, the authors described the different concepts that define e-Government according to the literature of the domain. Then, they explain and detail the benefits and challenges in its implementation. The authors outline that transforming public services is a long and difficult path that requires understanding of the existing situation, IT and functional skills, gathering skills, involving human resources and verifying continuously citizen satisfaction. Thus, transforming public services is by essence a long-term approach that requires support of the highest policy makers to ensure time, resources and financing and finally obtain tangible results.

The authors, ensured that the most possible perspectives are tackled from the needs of citizens and other stakeholders to the concrete implementation through cloud computing, architecture and security. However, the domain is closely linked to the evolution of information technology and constantly adjusts to new ideas and techniques. This justifies to highlight some new trends that already influence e-Government and requires a future update to remain relevant. The experience in developing digitalized public services proves that there is a tradeoff to define between securing and privacy of data when implementing innovative services [79]. The equilibrium should be defined according to the risks of the domain: it is easier to exchange data concerning companies and assets rather than the personal data of citizens.

The development of a proactive administration that offers automatic digital procedures currently faces limitations. First, the maturity of countries in classic e-Government is not necessarily sufficient. Second, digital procedures are often scattered in various websites that prevents IT workers to gather multiple information and achieve an end-to-end digital procedure. As a consequence, the initiative of the European Commission to establishing a single digital gateway that forces Member States to achieve 21 digitalized procedures (proof of residence, study grants, academic diplomas, vehicle registration, etc.) is a key effort to reach the required maturity [80] and move towards a-Government.

Another crucial orientation to hasten this development consists in introducing AI and deliver optimized public services. There are currently many experimentations that aim to propose new approaches that rely on generative AI and facilitate the job of civil servants. For example, Albert is a French project that helps them to answer questions of citizens [81]. As the authors are involved in public service transformation through their professional activities, this research should also be enriched by new experiences they encounter or imagine. This will provide additional feedbacks on how to transform public services in the coming years.

Disclaimer: This chapter represents the opinions of the authors and is the product of professional research. It is not meant to represent the position or opinions of their organizations.

References

1. UN. (2023). *Overview of e-Government*. United Nations. https://publicadministration.un.org/egovkb/en-us/Overview
2. Danziger, J. N., Kraemer, K. L., King, J. L. (1978). An assessment of computer technology in U.S. local governments. *Urban Systems, 3*(1), 21-37. ISSN 0147-8001, https://doi.org/10.1016/0147-8001(78)90004-9
3. Kraemer, K. L., et al. (1978). *Local government and information technology in the United States*. Paris: OECD Informatics Studies #12
4. Yildiz, M. (2007). e-Government research: Reviewing the literature, limitations, and ways forward. *Government Information Quarterly, 24*(3), 646–665. ISSN 0740–624X. https://doi.org/10.1016/j.giq.2007.01.002
5. Von Haldenwang, C. (2004). Electronic Government (e-Government) and Development. *European Journal of Development Research, 16*, 417–432. https://doi.org/10.1080/0957881042000220886
6. MacLean, D., & Titah, R. (2022). A systematic literature review of empirical research on the impacts of e-Government: a public value perspective. *Public Admin Rev, 82*, 23–38. https://doi.org/10.1111/puar.13413
7. Ariely, G. (2011). Why people (dis)like the public service: citizen perception of the public service and the NPM doctrine. *Politics & Policy, 39*, 997–1019. https://doi.org/10.1111/j.1747-1346.2011.00329.x
8. Chatfield, A. T. (2009). *Public Service reform through e-Government: a case study of 'e-Tax' in Japan 2009*, pp. 135–146. https://ro.uow.edu.au/infopapers/1712
9. Omar, K., Scheepers, H., & Stockdale, R. (2011). eGovernment service quality assessed through the public value lens. In *10th Electronic Government (EGOV)* (pp.431–440). Delft, Netherlands. https://doi.org/10.1007/978-3-642-22878-0_36
10. Huffman, B. D. (2017). E-Participation in the Philippines: A capabilities approach to socially inclusive governance. *JeDEM—EJournal of EDemocracy and Open Government, 9*(2), 24–46. https://doi.org/10.29379/jedem.v9i2.461.
11. Fang, Z. (2002). e-Government in digital era: Concept, practice, and development. *International Journal of the Computer, The Internet, and Management, 10*(2), 1–22. https://sahris.sahra.org.za/sites/default/files/additionaldocs/10.1.1.133.9080.pdf.
12. Liva, G., Codagnone, C., Misuraca, G., Gineikyte, V., & Barcevicius, E. (2020). Exploring digital government transformation: a literature review. In *Proceedings of the 13th International conference on theory and practice of electronic governance (ICEGOV '20)* (pp. 502–509). Association for Computing Machinery, New York, NY, USA. https://doi.org/10.1145/3428502.3428578
13. Grönlund, Å., & Horan, T. (2005). Introducing e-Gov: History, definitions, and issues. *Communications of the Association for Information Systems, 15*, pp-pp. https://doi.org/10.17705/1CAIS.01539.

14. Reitz, J. C. (2006). e-Government. *American Journal of Comparative Law, 54*, 2006. U Iowa Legal Studies Research Paper No. 05–43, Available at SSRN https://ssrn.com/abstract=887664.
15. Alshehri, M., & Drew, S. (2010). e-Government fundamentals. In *IADIS international conference ICT, society and human beings 2010*. https://hdl.handle.net/10072/37709.
16. Gupta, K.P., Singh, S., & Bhaskar, P. (2016). Citizen adoption of e-Government: a literature review and conceptual framework. *Electronic Government, an International Journal 12*(2), 160–185, https://doi.org/10.1504/EG.2016.076134
17. Malodia, S., Dhir, A., Mishra, M., & Bhatti, Z.A. (2021). Future of e-Government: An integrated conceptual framework. *Technological Forecasting and Social Change, 173*, 121102. ISSN 0040-1625. https://doi.org/10.1016/j.techfore.2021.121102.
18. Gioia, D. A., Corley, K. G., & Hamilton, A. L. (2013). Seeking qualitative rigor in inductive research: notes on the Gioia methodology. *Organizational Research Methods, 16*(1), 15–31. https://doi.org/10.1177/1094428112452151
19. Anthopoulos, L. G., Siozos, P., & Tsoukalas, I. A. (2007). Applying participatory design and collaboration in digital public services for discovering and re-designing e-Govern Applying participatory design and collaboration in digital public services for discovering and re-designing e-Government services. *Government Information Quarterly, 24*(2), 353–376. ISSN 0740–624X. https://doi.org/10.1016/j.giq.2006.07.018
20. Peristeras, V., Mentzas, G., Tarabanis, K.A., & Abecker, A. (2009). Transforming e-Government and E-participation through IT. *IEEE Intelligent Systems 24*, 5 (September 2009), 14–19. https://doi.org/10.1109/MIS.2009.103
21. Krishnan, S., Teo, T.S., & Lim, J. (2013). E-Participation and e-Government Maturity: A Global Perspective. In *International Working Conference on Transfer and Diffusion of IT (TDIT)* (pp. 420–435). Bangalore, India. https://doi.org/10.1007/978-3-642-38862-0_26
22. Steinbach, M., Sieweke, J., & Süß, S. (2019). The diffusion of e-participation in public administrations: A systematic literature review. *Journal of Organizational Computing and Electronic Commerce, 29*(2), 61–95. https://doi.org/10.1080/10919392.2019.1552749
23. Le Blanc, D. (2020). *E-participation: A quick overview of recent qualitative trends*. United Nations. https://doi.org/10.18356/0f898163-en
24. Heeks, R. (2001). *Understanding e-governance for development*. iGovernment (Working Paper no. 11). Available at SSRN https://ssrn.com/abstract=3540058 or https://doi.org/10.2139/ssrn.3540058.
25. Finger, M. (2003). From e-Government to e-Governance? Towards a model of e-Governance. In *Proceedings of the 3rd European conference on e government Eceg*. https://infoscience.epfl.ch/record/55629/files/frome-gov.pdf.
26. Rossel, P., & Finger, M. (2007). Conceptualizing e-Governance. In *Proceedings of the 1st international conference on Theory and practice of electronic governance (ICEGOV '07)* (pp. 399–407). Association for Computing Machinery, New York, NY, USA. https://doi.org/10.1145/1328057.1328141
27. Hasan, S. (2008). ICT, Internet, e-Government & E-governance: Concepts revisited. *The Chittagong University Journal of Social Science. 24*, 193–216. https://www.academia.edu/1347276/ICT_Internet_E_government_and_E_governance_Concepts_Revisited.
28. Dawes, S. S. (2008). The evolution and continuing challenges of e-governance. *Public Administration Review, 68*, S86–S102. https://doi.org/10.1111/j.1540-6210.2008.00981.x
29. Lee-Geiller, S., & Lee, T.D. (2019). Using government websites to enhance democratic E-governance: A conceptual model for evaluation. *Government Information Quarterly, 36*(2), 208–225. ISSN 0740–624X. https://doi.org/10.1016/j.giq.2019.01.003.
30. Irani, Z., Sahraoui, S., Özkan Yıldırım, S., Ghoneim, A., Elliman, T. (2021). T-government for benefit realization. In *presented at the 4th European and Mediterranean Conference on Information Systems, EMCIS 2007 (24–26 Haziran 2007)*, Valencia, Spain. Accessed: 00, 2021. [Online]. Available https://www.scopus.com/inward/record.uri?partnerID=HzOxMe3b&scp=84857480742&origin=inward.
31. King, S., & Cotterill, S. (2007). Transformational Government? The role of information technology in delivering citizen-centric local public services. *Local Government Studies, 33*(3), 333–354. https://doi.org/10.1080/03003930701289430

32. Janssen, M. and Shu, W.S. (2008). Transformational government: basics and key issues. In *Proceedings of the 2nd international conference on Theory and practice of electronic governance (ICEGOV '08)* (pp. 117–122). Association for Computing Machinery, New York, NY, USA. https://doi.org/10.1145/1509096.1509120
33. Omar, A., Weerakkody, V., Daowd, A. (2020). Studying transformational government: A review of the existing methodological approaches and future outlook. *Government Information Quarterly, 37*(2). Article 101458. https://doi.org/10.1016/j.giq.2020.101458
34. Björses, M. and Ivarsson, A. (2019). *The dynamic balancing act Enacting ambidextrous IT Governance within the public sector*. Master theses. http://hdl.handle.net/2077/61411
35. Ndlovu, N., Ochara, N.M., & Martin, R. (2022). Influence of digital government innovation on transformational government in resource-constrained contexts. *Journal of Science and Technology Policy Management*, ahead-of-print (ahead-of-print). https://doi.org/10.1108/JSTPM-11-2021-0173.
36. Karunasena, K., Deng, H., & Singh, M. (2011). Measuring the public value of e-Government: A case study from Sri Lanka. *Transforming Government: People, Process and Policy, 5*(1), 81–99. https://doi.org/10.1108/17506161111114671
37. Mensah, I. K., Zeng, G., & Mwakapesa, D. S. (2022). Understanding the drivers of the public value of e-Government: Validation of a public value e-Government adoption model. *Frontiers in Psychology, 13*, 962615. https://doi.org/10.3389/fpsyg.2022.962615
38. Castelnovo, W. (2013). A stakeholder based approach to public value. In *Proceedings of the 13th European Conference on Egovernment, ECEG 2013* (pp. 94–101).
39. Boyd, S., & Vandenberghe, L. (2004). *Convex optimization*. Cambridge University Press.
40. Gaie, C., Assaad, M., Muck, M., & Duhamel, P. (2008). Distributed discrete resource optimization in heterogeneous networks. In *2008 IEEE 9th workshop on signal processing advances in wireless communications* (pp. 560–564). Recife, Brazil. https://doi.org/10.1109/SPAWC.2008.4641670
41. Gaie, C. (2010). *Allocation de ressources à prise de décisions distribuées dans des réseaux mobiles hétérogènes*. Informatique. Paris Saclay. Français, https://hal.science/tel-02884997v1/document
42. Meiyanti, R., Utomo, B., Sensuse, D. I., & Wahyuni, R. (2018). e-Government challenges in developing countries: a literature review. In 2018 6th International Conference on Cyber and IT Service Management (CITSM) (pp. 1–6). Parapat, Indonesia. https://doi.org/10.1109/CITSM.2018.8674245
43. Aldosemani, T. I. (2023). e-school initiatives that instigated digital transformation in education: a case study according to SABER-ICT framework. In C. Gaie, M. Mehta, (Eds.), Recent advances in data and algorithms for e-Government. artificial intelligence-enhanced software and systems engineering, vol. 5. Springer. https://doi.org/10.1007/978-3-031-22408-9_2.
44. Machmud, M., Musa, A. E. Z., Suprapto, B., Salahudin. (2023). The role of public libraries in improving public literacy through twitter social media in Indonesia. In C. Gaie, & M. Mehta, (Eds.), *Recent advances in data and algorithms for e-Government. artificial intelligence-enhanced software and systems engineering*, vol. 5. Springer. https://doi.org/10.1007/978-3-031-22408-9_10
45. Bhuvana, M., Ramkumar, A., & Neeraja, B. (2023). Machine learning technique for predicting the rural citizens' trust on using e-governance health care applications during COVID-19. In C. Gaie, M. Mehta, (Eds.), *Recent advances in data and algorithms for e-Government. Artificial intelligence-enhanced software and systems engineering*, vol. 5. Springer. https://doi.org/10.1007/978-3-031-22408-9_6
46. Gaie, C., Mueck, M. (2023). Enhancing government actions against Covid-19 using computer science. In C. Gaie, M. Mehta, (Eds.), *Recent advances in data and algorithms for e-Government. Artificial intelligence-enhanced software and systems engineering,* vol. 5. Springer. https://doi.org/10.1007/978-3-031-22408-9_8
47. Gkikas, D. C., Theodoridis, P. K., Gkikas, M. C. (2023). Artificial Intelligence (AI) Use for e-governance in agriculture: exploring the bioeconomy landscape. In C. Gaie, M. Mehta (Eds.), *Recent advances in data and algorithms for e-Government. artificial intelligence-enhanced*

software and systems engineering, vol 5. Springer, Cham. https://doi.org/10.1007/978-3-031-22408-9_7
48. Gaie, C. (2023). Struggling against tax fraud, a holistic approach using artificial intelligence. In C. Gaie, M. Mehta (Eds.), *Recent advances in data and algorithms for e-Government. artificial intelligence-enhanced software and systems engineering*, vol 5. Springer, Cham. https://doi.org/10.1007/978-3-031-22408-9_4
49. Arief, A., Ayub Wahab, I.H., & Muhammad, M. (2021). Barriers and challenges of e-Government services: a systematic literature review and meta-analyses. *IOP Conference Series: Materials Science and Engineering, 1125*. https://doi.org/10.1088/1757-899x/1125/1/012027
50. Heidlund, M., & Sundberg, L. (2022). Evaluating e-Government: Themes, trends, and directions for future research. *First Monday, 27*(12). https://doi.org/10.5210/fm.v27i12.12526
51. Efe, A., Kazdal, H. (2019). IT Security trends for e-Government threats. *International Journal of Multidisciplinary Studies and Innovative Technologies*, 3(2), 105–110. https://dergipark.org.tr/en/pub/ijmsit/issue/50263/403090
52. Yang, L., Elisa, N., & Eliot, N. (2019). Chapter 7—Privacy and security aspects of e-Government in smart cities. In: D. B. Rawat & K. Z. Ghafoor, *Smart cities cybersecurity and privacy* (pp. 89–102). Elsevier. ISBN 9780128150320. https://doi.org/10.1016/B978-0-12-815032-0.00007-X
53. [Khayyat2020] Khayyat, M.; Alhemdi, F.; Alnunu, R. The Challenges and Benefits of Blockchain in e-Government. Int. J. Comput. Sci. Netw. Secur. 2020, 20, 15–20. http://paper.ijcsns.org/07_book/202004/20200403.pdf
54. Kassen, M. (2022). Blockchain and e-Government innovation: Automation of public information processes. *Inf Syst 103*, C (Jan 2022). https://doi.org/10.1016/j.is.2021.101862
55. Wang, S., & Sun, C. (2023). *Blockchain empowered dynamic spectrum sharing: Standards, state of research and road ahead*. TechRxiv. Preprint. https://doi.org/10.36227/techrxiv.19395176.v1
56. Medaglia, R., & Damsgaard, J. (2020). Blockchain and the United Nations sustainable development goals: towards an agenda for IS research. In D. Vogel, K. N. Shen, & P. S. Ling (Eds.), *PACIS 2020 Proceedings [36] Association for Information Systems. AIS Electronic Library (AISeL)*. http://aisel.aisnet.org/cgi/viewcontent.cgi?article=1035&context=pacis2020
57. Gao, Y., Pan, Q., Liu, Y., Lin, H., Chen, Y., & Wen, Q. (2021). The notarial office in e-Government: a blockchain-based solution. *IEEE Access, 9*, 44411–44425. https://doi.org/10.1109/ACCESS.2021.3066184
58. Al Mudawi, N., Beloff, N., & White, M. (2020). Issues and challenges: Cloud computing e-Government in developing countries. *International Journal of Advanced Computer Science and Applications (IJACSA), 11*(4). https://doi.org/10.14569/IJACSA.2020.0110402
59. Suleiman, M., Matinja, Z., Musa, Z., & Abdulkadir, Z. (2020). Cloud computing is an integral tool for e-Government: challenges and prospects. IX. pp. 2824–2836. https://www.researchgate.net/publication/342479897_Cloud_Computing_Is_An_Integral_Tool_For_e-Government_Challenges_And_Prospects
60. Fathey, M., Olayah, F., Ali, A., & Gazem, N. A. (2020). The effect of cloud computing adoption on the sustainability of e-Government services: a review. *International Journal of Advanced Science and Technology, 29*(05), 2636–2642. http://sersc.org/journals/index.php/IJAST/article/view/11161
61. Al-Shargabi, B., Al-Jawarneh, S., Hayajneh, S. M. A. (2020). A cloudlet based security and trust model for e-Government web services. *Journal of Theoretical and Applied Information Technology, 98*(1), 27–37 Posted. http://www.jatit.org/volumes/Vol98No1/3Vol98No1.pdf
62. Akbar, P., Nurmandi, A., Irawan, B., & Jafar Loilatu, M. (2022). Research trends in e-Government interoperability: mapping themes and concepts based on the scopus database. *JeDEM—EJournal of EDemocracy and Open Government, 14*(2), 83–108. https://doi.org/10.29379/jedem.v14i2.707
63. Saputro, R., Pappel, I., Vainsalu, H., Lips, S., & Draheim, D. (2020). Prerequisites for the adoption of the X—road interoperability and data exchange framework: a comparative study. In 2020 Seventh international conference on edemocracy & egovernment (ICEDEG), Buenos Aires, Argentina, pp. 216–222. https://doi.org/10.1109/ICEDEG48599.2020.9096704

64. Campmas, A., Iacob, N., & Simonelli, F. (2022). How can interoperability stimulate the use of digital public services? An analysis of national interoperability frameworks and e-Government in the European Union. *Data & Policy, 4*, E19. https://doi.org/10.1017/dap.2022.11
65. Puspitasari, N., Budiman, E., Sulaiman, Y. N., Firdaus, M. B. (2021). "Microservice Api Implementation For e-Government Service Interoperability. *J Phys Conf Ser., 1807*(1), 012005. https://doi.org/10.1088/1742-6596/1807/1/012005
66. Gaie. (2021). An API-intermediation system to facilitate data circulation for public services: the French case study, Christophe Gaie. *International Journal of Computational Systems Engineering* 6(4), 201–210. https://doi.org/10.1504/IJCSYSE.2021.120292
67. Vaccari, L., Posada, M., Boyd, M., & Santoro, M. (2021). APIs for EU governments: A landscape analysis on policy instruments, standards. *Strategies and Best Practices. Data, 6*, 59. https://doi.org/10.3390/data6060059
68. Berrada, M., & El Akkad, N. (2019). Artificial intelligence based composition for e-Government services. In Proceedings of the third international conference on computing and wireless communication systems, ICCWCS 2019, Faculty of Sciences, Ibn Tofaïl University-Kénitra-Morocco. https://doi.org/10.4108/eai.24-4-2019.2284071
69. Al-Mushayt, O. S. (2019). Automating e-Government services with artificial intelligence. *IEEE Access, 7*, 146821–146829. https://doi.org/10.1109/ACCESS.2019.2946204
70. Zuiderwijk, A., Chen, Y.-C., & Salem, F. (2021). Implications of the use of artificial intelligence in public governance: a systematic literature review and a research agenda. *Government Information Quarterly*. https://doi.org/10.1016/j.giq.2021.101577
71. Yerlikaya, S., Erzurumlu, Y.Ö. (2021). Artificial intelligence in public sector: a framework to address opportunities and challenges. In A. Hamdan, A. E. Hassanien, A. Razzaque, B. Alareeni, (Eds.), The *fourth industrial revolution: implementation of artificial intelligence for growing business success. Studies in computational intelligence*, vol. 935. Springer, Cham. https://doi.org/10.1007/978-3-030-62796-6_11
72. Gaie, C., & Mueck, M. (2022). *An artificial intelligence framework to ensure a tradeoff between sanitary and economic perspectives during the COVID-19 pandemic, deep learning for medical applications with unique data.* https://doi.org/10.1016/B978-0-12-824145-5.00008-3
73. Barthelmess, B., & Langlois-Berthelot, J. (2022). How information technologies are changing forensic finance in Africa. *Africa Trends, 9*(2). https://idsa.in/system/files/page/2015/africa trend_9_2.pdf
74. Karyono, O., & Agustina, K. (2019). Determining the priority strategy in the implementation of e-Government through Swot analysis model. *Bp Int Res Crit Inst Humanit Soc Sci 2*, 66–74. https://doi.org/10.33258/birci.v2i2.238
75. Gacitúa, R., Astudillo, H., Hitpass, B., Osorio-Sanabria, M., & Taramasco, C. (2021). Recent models for collaborative e-Government processes: a survey. *IEEE Access, 9*, 19602–19618. https://doi.org/10.1109/ACCESS.2021.3050151
76. Halsbenning, S., Niemann, M., Distel, B., Becker, J. (2021). Playing (Government) Seriously: Design Principles for e-Government simulation game platforms. In Ahlemann, F., Schütte, R., Stieglitz, S. (Eds.), *Innovation through information systems. WI 2021. Lecture notes in information systems and organisation*, vol 48. Springer, Cham. https://doi.org/10.1007/978-3-030-86800-0_6
77. Anshari, M., & Hamdan, M. (2022). Enhancing e-Government with a digital twin for innovation management. *Journal of Science and Technology Policy Management, ahead-of-print* (ahead-of-print). https://doi.org/10.1108/JSTPM-11-2021-0176
78. Frogeri, R. F., da Silva Diniz, W. F., dos Santos Portugal Júnior, P., Piurcosky, F. P. (2023). e-Government and green IT: The intersection point. In C. Gaie, M. Mehta, (Eds.), *Recent advances in data and algorithms for e-Government. artificial intelligence-enhanced software and systems engineering*, vol 5. Springer, Cham. https://doi.org/10.1007/978-3-031-22408-9_5
79. Valkenburg, G. (2015). Privacy Versus Security: Problems and Possibilities for the Trade-Off Model. In S. Gutwirth, R. Leenes, P. de Hert, (Eds.), *Reforming european data protection law. Law, governance and technology series*, vol 20. Springer, Dordrecht. https://doi.org/10.1007/978-94-017-9385-8_10

80. EU. (2018). Regulation (EU) 2018/1724 of the European Parliament and of the Council of 2 October 2018 establishing a single digital gateway to provide access to information, to procedures and to assistance and problem-solving services and amending Regulation (EU) No 1024/2012 (Text with EEA relevance.). PE/41/2018/REV/2. http://data.europa.eu/eli/reg/2018/1724/oj.
81. CNIL. (2023). https://www.cnil.fr/en/artificial-intelligence-and-public-services-sandbox-cnil-supports-8-innovative-projects.
82. Davison, R. M., Wagner, C., & Ma, L. C. K. (2005). From government to e-Government: A transition model. *Information Technology & People, 18*(3), 280–299. https://doi.org/10.1108/09593840510615888
83. Huffman, B. D. (2017). E-Participation in the Philippines: A Capabilities Approach to Socially Inclusive Governance. *JeDEM—EJournal of EDemocracy and Open Government, 9*(2), 24–46. https://doi.org/10.29379/jedem.v9i2.461
84. [Elisa2020] Elisa, N., Yang, L., Li, H., Chao, F., & Naik, N. (2020). Consortium Blockchain for Security and Privacy-Preserving in e-Government Systems. https://arxiv.org/abs/2006.14234

Chapter 2
Enhancing Citizen Satisfaction Using Citizen-Facing Process Mining

Shreekanth M. Prabhu, Natarajan Subramanyam, and V. Jagadishwari

Abstract Processes are at the heart of day-to-day reality for citizens. These include processes where citizens have to interact with government institutions as well as processes where citizens have to engage with public institutions such as banks, insurance companies, and online platforms. It is generally observed that many well-intentioned policies and systems when operationalized as processes fail to achieve their goals. To make it worse, even some basic processes can cause a lot of anguish to the citizens and the government officials/leaders of public institutions may remain blissfully unaware of ground reality. Automation is often cited as a silver bullet to address citizen concerns. But automation works purposefully only if processes can be monitored and enhanced when in motion. In this context, the emerging field of process mining holds great promise. In this chapter, we focus on citizen-facing process mining. Here the citizen perspective and how a process is practiced in the field becomes a lot more important. We propose an architectural framework for citizen-facing process mining and a research model that can be used to enhance citizen satisfaction and delight citizens.

Keywords Citizen satisfaction · Process mining · Architectural framework · Research model

---

These authors contributed equally to this work.

S. M. Prabhu (✉) · V. Jagadishwari
Department of Computer Science and Engineering, CMR Institute of Technology, Bengaluru, Karnataka, India
e-mail: shreekanthpm@gmail.com

N. Subramanyam
Department of Computer Science and Engineering, PES University, Bengaluru, Karnataka, India

C. Gaie and M. Mehta (eds.), *Transforming Public Services—Combining Data and Algorithms to Fulfil Citizen's Expectations*, Intelligent Systems Reference Library 252, https://doi.org/10.1007/978-3-031-55575-6_2

2.1 Introduction

Citizen satisfaction is a worthy goal for institutions and Governments to pursue. However, at any point in time, it tends to be elusive and rather a moving target. Many cases citizen dissatisfaction stems from (i) apathy and bureaucratic mind-set of workers (ii) lack of initiative and interest of officials who are supposed to oversee departments and functions (iii) frequent shuttling of a given case between departments and between institutions and citizens. Further, the sheer complexity of the job at hand may mean simplifying things can lead to fraud. In many cases, the institutions respond to these challenges by automating the processes and workflows. This works well in many cases that are routine. But in certain cases, it can lead to reduced citizen satisfaction as well as a larger number of frauds.

Citizens need to interface with Government Departments and Public Institutions, nearly on a day-to-day basis. The government is involved in certain processes such as law and order, traffic management, and issuing licenses and permits that can be categorized as governance processes. Many government institutions such as Police and Tax authorities may consider themselves to be part of the ruling establishment and may not bother about citizen satisfaction. They may feel citizens do not have a choice but to approach them.

Then the Governments are also involved in development and welfare activities. Here development activities are laying a new road, starting a new school or hospital, and so on. The welfare activities generally comprise providing benefits such as old age pensions, PDS (Public Distribution System) grains, and varied schemes that facilitate treatment of people under health insurance/health assurance schemes and so on. Here the Government officials may have a mindset that they are dispensing favours. Thus, citizen satisfaction can easily go on the back burner.

Even when Governments are well-intentioned, they may run into obstacles. To materialize their good intention, Governments formulate appropriate policies. The policies then are incorporated into legislations and rules and finally are operationalized as processes. To give an example, the policy on towing vehicles is intended to avoid impediments to a smooth flow of traffic at critical sections. This also presupposes that there is a clear demarcation of tow-away zones and the personnel who tow do it in a way that meets the intent of the policy and spirit of the legislation. This is particularly important for coercive actions such as towing that can cause angst and trauma to citizens. But some officials tow vehicles even when the traffic is lean when it is easier to tow them as they view towing as a revenue-earning operation for the government, rather than seeing it as a law enforcement operation in the public interest.

A more pervasive and critical Government function however is that of regulation. Here Government lays down laws and guidelines on how citizens need to be authenticated, and how citizens and institutions transact among themselves to exchange goods and services, and how they engage in operations such as payments. Here Governments provide legal infrastructure as well as technology framework as exemplified by the UPI framework in India for digital payments. Here again, mechanisms that work well in certain situations may turn out to be counter-productive in other situations. For example, Governments typically facilitate a framework to authenticate people and their addresses and facilitate mutual payments. In India, Aadhaar eSign is a service that can authenticate individuals who consent to update information or register a claim. This is used in lieu of physical signatures in manual systems. This works fine if names match perfectly and OTP (One Time Password) is generated on an Aadhaar-linked mobile phone, which serves as a private key. Here legacy issues can come into play, as names in the Indian context tend to get recorded in a varied manner, at times with middle names, without middle names, abbreviated middle names, and certain names can be spelled in multiple ways and so on. In such cases, a lack of clear alternative authentication processes, whenever eSign fails, can cause enormous anguish to citizens. In many cases, citizens do not have information on whom to communicate as multiple agencies are involved and those whom they communicate may not respond.

Considerations of security and privacy mean there should be caution while making processes simple, transparent, and speedy. In India, bank accounts are linked with mobile numbers of citizens, and use One-Time-Password for authentication. Here when a subscriber changes mobile SIM, a cooling period of 24 hours is mandated, to avoid SIM change fraud. Here a citizen who is not well-informed may expect faster service Then there are sectors such as banking and insurance where both public sector units owned by government and private institutions participate. Typically, when it comes to Government institutions, it is generally hard to enroll in service and equally hard to leave it. However private institutions may make it easy to subscribe but extremely difficult to leave a service. These are typically services where they have to solicit customers. Then there are also many unethical practices where a citizen's consent to subscribe to a service is assumed to be perpetually valid and subscriptions get auto-renewed with charges accrued to their credit cards unless citizens remember to and choose to cancel a service. Such practices cause financial loss to citizens who are not mindful.

Then there are sectors where multiple institutions follow the same or similar processes and use common software platforms/packages. Here they assume a certain familiarity with nomenclature among citizens. In one instance, to set the first (green) PIN (Personal Identification Number) for an ATM Card, a citizen is expected to enter their date of birth in "ddmmyy" format using an ATM Machine. This seemingly simple requirement can stump people who are not IT savvy. Thus, a proliferation of such citizen stories makes a case for citizen-facing process mining which captures citizens' perspectives. There are also happy citizen stories. In one case, a bank official reduced the friction in this process, by engaging a helper to assist citizens in setting their green PIN. Classical/Institutional Process Mining makes use of event logs as input and then deals with process discovery where a process model is discovered. Then it is used to check process conformance and look for ehancement opportunities. There are control flow views, organization views, and case views which are used in many case studies. Petri net is a popular approach to model processes. The less explored view is the information view. This we term as information system view, which is particularly important as a majority of processes are becoming increasingly digital. With the information system view, the design of forms and user interfaces becomes an important aspect. Another related opportunity is cross-institutional process mining. This is particularly useful when there is a commonality in process steps across institutions and they leverage common/similar infrastructure/practices/software packages.

In this work, we make use of concepts and precepts from Process Mining. That however is more as a back-drop. The thrust of this paper is to address architectural concerns in monitoring citizen-facing processes and operationalizing the processes better. To this end, we propose an Architectural Framework for Citizen-Facing Process Mining and a Research Model to enhance citizen satisfaction. Unlike Process Mining which makes use of event logs as inputs, Citizen-Facing Process Mining works with Citizen Stories and arrives at action plans and routines to follow to enhance citizen satisfaction.

In Citizen-Facing Process Mining, we need to look at processes from a citizen's viewpoint. For a citizen, the process is not done, till his job is done. This is unlike an institution's view with a control flow perspective where a control flow ends in a good state or an error. In citizen-facing processes, preventing fraud is also equally important. The rest of the chapter is as follows. Section 2.2, Citizen Satisfaction, covers the literature on Citizen Satisfaction. Section 2.3, Process Mining gives an overview of Process Mining and Applications. Section 2.4, Architectural Framework to Manage Citizen Satisfaction dwells on architectural concerns related to Citizen-Facing Processes. Section 2.5, Research Model proposes a new model to research citizen satisfaction on citizen-facing processes. Section 2.6 covers an analysis of Citizen Stories based on the proposed research model. Section 2.7 Discussions summarize insights gained and propose an action plan to operationalize Citizen Satisfaction Research. Section 2.8 Conclusions, concludes the chapter.

2.2 Citizen Satisfaction

Over the last few decades, there have been numerous studies on Consumer/Customer Satisfaction and its relationship with service quality. Generally, it may be considered a given that judgment of service quality is a causal antecedent for judgment of customer satisfaction, However, even the opposite is found to be true in certain research studies. Parasuraman [1–4] has done pioneering research in understanding the inter-relationship between service quality and customer satisfaction. Oliver [5, 6] analysed consumer satisfaction from a variety of perspectives.

Citizen Satisfaction is a broader concept than Customer Satisfaction. When he consumes a public service, a citizen indeed is a customer. A citizen can also be a beneficiary when she avails benefits of Government policies and a sponsor as she funds the Government's expenses through taxes. Certain public services and duties Governments have to do mandatorily. They also need to be more considerate and fairer. In general, public service is viewed by citizens using an ethical/moral prism. Thus, he expects the Government not to be business-like and from his side may be willing to put up with a degree of inconvenience for a good cause. Andrews and

Table 2.1 Seoul service quality index

Dimension	Components
Process quality	Speed, credibility and kindness
Outcome quality	Satisfaction, likeability
Office environment quality	Amenity, convenience and aesthetics
Social quality	Consideration for socially weak groups, security and privacy

Boyne [7, 8] have done insightful work related to citizen satisfaction with public services. Andrews and Boyne [9], and Wichowsky and Moynihan [10] refer to the dual role citizen plays. The citizen as a citizen is influenced by moral ideals and as a customer influenced by performance considerations.

Reddick [11] compares three alternative theories pertaining to citizen satisfaction with public services. Reform theory argues that citizens derive the most satisfaction from larger government units because of economies of scale in service delivery. Political economy theory argues that citizens achieve the most satisfaction from smaller local government units since their voice is more likely to be heard. The third theory argues that citizens have different levels of satisfaction depending upon the specific services received and directly experienced. The third theory is pursued in this paper, while being mindful of the broader context and landscape.

There are important differences in the way public institutions and private institutions operate. In a private institution, outcomes are generally more important than processes. In a public institution processes become more important than outcomes. This is natural as any actions taken in public institutions are open to legal challenge. Even privately run banks can be treated as public institutions as they are answerable to regulators. Thus, it requires really conscientious individuals to go against the tide. Kelly [12] discusses the limited effect of administrative performance measures on citizen satisfaction and concludes that citizen Satisfaction as a whole is situational and contextual. However, it is not always the case that the private sector outperforms the public sector. A study by Mengi [13] showed that there is higher citizen satisfaction with PSU (Public Sector Unit) banks than Private Sector Banks. This goes against the assumption that privatization improves service quality and citizen satisfaction.

Im and Lee [14] studied citizen satisfaction with local management in Seoul City. They defined the Seoul Service Quality Index that is given in Table 2.1. This is followed by the Seoul Service Potential Index given in Table 2.2. Their findings indicate the quality of management has a positive effect on the quality of service.

In summary, Citizen Satisfaction turns out to be a complex phenomenon that combines expectations, experiences, and previous attitudes. It is also a relative indicator where a citizen compares his current experience with his earlier experience, which would have set certain expectations. The prior experience could be with the same institution or with a totally different institution. In particular, citizens experience far

Table 2.2 Seoul service potential index

Dimension	Components
Leadership	Leadership strategies and planning
Resources	Human resources, knowledge resources, material resources and budget
Relationships	Citizen relationship, partner firm relationship, community relationships
Process process management	Process improvement

greater happiness when they are satisfied with public services, as in some sense they have a sense of association, belonging, or ownership that is far greater than with the private sector. High-quality public service also fosters citizen confidence and trust. Various facets of citizen satisfaction are covered by Van de Walle [15], Bradford [16] and Cohen [17].

Citizen satisfaction is also a matter of judgment. It is also about the opinions and feelings of citizens. In institutions where a citizen is treated with respect and care, some degree of inefficiency and delay may be acceptable. Citizen satisfaction also depends on expectations. These expectations vary from citizen to citizen and they are shaped by a variety of factors such as their past experience, what they have heard from their acquaintances as well as general perception in media which includes social media. There are three approaches used to study citizen satisfaction: the Disconfirmation Theory, the Service Quality Model, and the Performance Model.

According to disconfirmation theory, satisfaction depends on relative performance compared to what citizen expects. If a citizen expects a bad service, then even normal service may delight him. On the flip side, if a citizen expects very good service, even a normal service can disappoint him. The citizen's expectation depends on his prior experience and his social learning. Expectations can be categorized in two ways: (1) positive expectations and (2) normative expectations. In simple words what citizen hopes for is considered a positive expectation and what can be normally and realistically be expected is normative expectation. Chatterjee and Suy [18] discuss disconfirmation theory in the context of public services.

The second theory predicates that citizen satisfaction depends on service quality. Parasuraman et al. [2] in their highly cited work, identify ten parameters critical to service quality, namely Reliability, Responsiveness, Competence, Accessibility, Courtesy, Communication, Credibility, Security, Understanding, and Tangibility. Later research in this direction boiled down these parameters to five. Table 2.3 gives a generic definition of the five parameters. This model is called SERVQUAL. There are several studies on customer/consumer/citizen satisfaction in the banking sector that consider Service Quality as an antecedent to Citizen Satisfaction (Table 2.4). A study by Mohammad [19] covers the commercial banking sector in Jordan. Mistry [20] did a study with special reference to banks in Surat City, India. Table 2.5 describes how Mistry applies these five parameters to the banking sector. Singh and Kaur [21] did a study focused on banks in Chandigarh, which covered 3 additional

Table 2.3 Service quality parameters

Sr No.	Parameter	Definition
1	Reliability	Ability to deliver the promised service dependably and with accuracy
2	Responsiveness	Willingness to support customers and deliver prompt service
3	Assurance	Knowledge and courtesy towards employee and their ability to inspire trust/confidence
4	Empathy	Caring, individualization and attention closeness
5	Tangibles	Appearance of personnel, physical facilities and equipment

Table 2.4 Service quality parameters applied to banking sector

Sr No.	Parameter	Explanation
1	Reliability	Following up on promises in timely manner; Error-free records and when a problem is reported bank shows sincere interest in solving it
2	Responsiveness	Employees give prompt service; Employees are always willing to help; Employees will tell exactly when the service will be performed
3	Assurance	The behaviour of employees instils confidence; transaction safety; consistent courtesy; employees knowledgeable to answer questions
4	Empathy	Understanding specific needs of citizens; operating hours convenient to citizens; individual attention
5	Tangible	Modern looking equipment; presentable employees; satisfactory physical facilities and amenities

factors: word of mouth, social responsibility, and service innovation. Fida et al. [22] did a study on the banking sector in Oman. Oman Bank showed that empathy and responsiveness dimensions have a significant positive impact on citizen satisfaction. Jamal and Naser [23] did a study on customer satisfaction in Abu Dhabi Commercial Bank that emphasized the importance of covering both core and relational dimensions of service quality. In particular, in the case of banks, access hours become very important, as most banks are not open when citizens are having their off-hours. If you compare this with the Telecom Service Providers in India, their access hours stretch into the evenings. The quality of a service can be assessed in two steps–(1) the way the service is rendered and (2) the result of the rendered service. Gronroos [24] has identified two areas of service quality namely—technical quality and functional quality. Technical quality is concerned with what the consumer receives while functional quality emphasizes the process (how).

Oliver [6] argued that satisfaction is a function of quality. Satisfaction increases as service quality surges. Engdaw [25] claimed satisfaction is generally high as long as quality service delivery is maintained. According to some researchers, customer satisfaction is short-term as it is a service-specific experience whereas the judgment of quality is long-term because it's a cognitive process. Thus, service quality is a useful lever for institutions to work on.

Table 2.5 Performance model

Sr No.	Parameter	Definition
1	Citizen satisfaction	Fully satisfied, satisfied to some extent, dissatisfied to some extent, fully dissatisfied, not accessed service don't know/can't say
2	Compliance responsiveness timely delivery Services own self (without third party assistance)	Strongly agree, agree to some extent, disagree to some extent strongly disagree, don't know/can't say
3	Paid/Was asked bribe	Yes, no, did not receive any public service, don't know/can't say
4	Service attempts	Number of attempts

The third model is the performance model, where citizen satisfaction is an integral part. Here, it is about judging say how a government is working compared to how it should work. This takes a rather macro view, which looks at satisfaction, quality, ethical issues, and friction. Performance Model is implemented by deploying surveys to get feedback from citizens. Compared to taking surveys on a narrow service or a single product, these surveys need more care, as only a fraction of citizens may access a fraction of the services. Table 2.5, illustrates the operationalization of the performance model using a survey design adopted by Lamsal and Gupta [26].

The satisfaction surveys will also be influenced by information. The information is particularly important for those services that are not directly experienced by citizens or used very frequently. According to James and Wilson [27], information about excellent performance would increase citizen satisfaction whereas satisfaction would decrease with the information of poor performance.

In the last few decades, the bulk of services have become digital and online. Thus, we need to bring information systems into the ambit of citizen satisfaction and service quality. The accuracy and quality of information is another critical factor for banks. A message from the bank with sketchy details can easily startle citizens and a lack of communication with updates, warnings, and alerts is equally worrying. Making services online makes it easier to make the processes transparent.

The study by Yang [28] on China's healthcare system underscores the importance of transparency as an enabler for citizen satisfaction. Three facets of transparency are disclosure, clarity, and accuracy. In public institutions, transparency is the availability of information to the general public and clarity about government rules, regulations, and decisions. Transparency ensures that citizens' expectations are realistic and rational. Greater transparency also increases trust. Transparency reduces information asymmetry between citizens and institutions. However, Yang's study determined that perceived transparency is a double-edged sword, as increased transparency raises expectations for service quality, which subsequently dampens the positive effects of service quality and perceived value on citizen satisfaction.

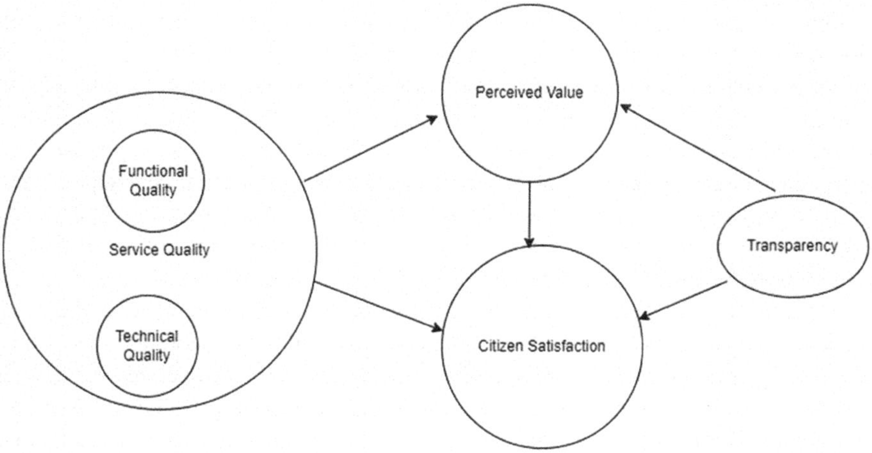

Fig. 2.1 Extended service quality framework

Figure 2.1 depicts the extended framework for the relationship between Service Quality and Citizen Satisfaction. Here, Service Quality has two components. Functional Quality caters to the quality of the process used to provide service. Technical Quality is the quality of service received. Here perceived transparency influences the perception of value as well as citizen satisfaction. In general, service quality is the antecedent and citizen satisfaction is consequent, the perceived value of the service itself is a mediating factor (Fig. 2.1). The perceived value depends on sacrifices like price and non-monetary costs. The value is generally paid less attention, as most policies are designed for the public at large and not specific citizens. The skills of citizens in participating in the co-creation of services i.e., doing their part with competence also influence their satisfaction. Situational and personal factors also influence citizen satisfaction. Considering that service quality is critical to citizen satisfaction. The following questions are very pertinent:

1. What is the role of the core, relational, and tangible dimensions of service quality in the formation process leading to overall citizen satisfaction?
2. What is the role of citizen expertise in moderating the effects of various dimensions of service quality in the formation process leading to overall citizen satisfaction?
3. What is the role of citizen differences such as demographics and factors such as ethnic background or length of stay in moderating the effect of service quality in the formation process leading to overall citizen satisfaction?

Many institutions, in particular those in the insurance industry assess service quality using complaints ratio [29]. They miss out on the feelings and perceptions of those who do not complain as well as specific knowledge about service shortfalls that led to the complaints.

Some scholars link citizen satisfaction to events and encounters and interpret good service quality as consistency in delivering that satisfaction. Among the service quality parameters, it is reliability that requires longitudinal studies. Ensuring high reliability calls for giving reliability a premium consideration during recruitment and training. This should be accompanied by sustained oversight by supervisors to ensure that reliability becomes the norm. Supervisors should particularly pay heed to out-of-ordinary or eccentric requests, which elude standard operating procedures and then intervene as needed. Institutions should also communicate that reliability is their strong suit to the stakeholders.

For high customer satisfaction, the interventions required may be at the system level. These include executive management's ability to encourage and motivate employees to act in a citizen-friendly manner, ensuring adequate resources and capacity as well as ensuring good internal/external relationships to ensure transactions happen within reasonable time, and removing impediments such as bureaucratic inertia and administrative rigidity.

2.3 Process Mining

Process Mining is a combination of Data Science and Process Science. Data Science revolves around data and Process science is model-centric. Process mining views these two domains as complementary to each other and harnesses the combined outcome from Data Science and Process Science. In Process mining, event data is used to discover process models. The discovered models are used to check the conformance of the model and also to propose enhancements to the models. Process mining caters to Process Discovery, Conformance, and Process Enhancement as depicted in Fig. 2.2.

Input to process mining is a log of activities. The log pertains to a set of cases. There are a group of events [30] associated with each case. Each case is represented using a case ID and events are represented with unique event IDs. A sample log is shown in Table 2.6. This log data is from an automobile insurance company, where people insure their vehicles and claim compensation in the event of an accident. There are six unique activities that a case can go through namely First Notification of Loss (FNOL), assign claim, claim decision, payment sent, set reserve, and close claim. It has five properties the time stamp, the name of the activity, the name of the agent handling the activity, the name of the claimant, and the compensation amount that has been claimed. There can be more properties like the age of the claimant, address of the claimant, etc. The event log shows two entries, with two cases having the case IDs as Case 1 and Case 2. Each case consists of three events. Row one of Case 1 indicates that a First Notification of Loss (FNOL) was filed by claimant Dale with agent Pete claiming a compensation of 12,345.

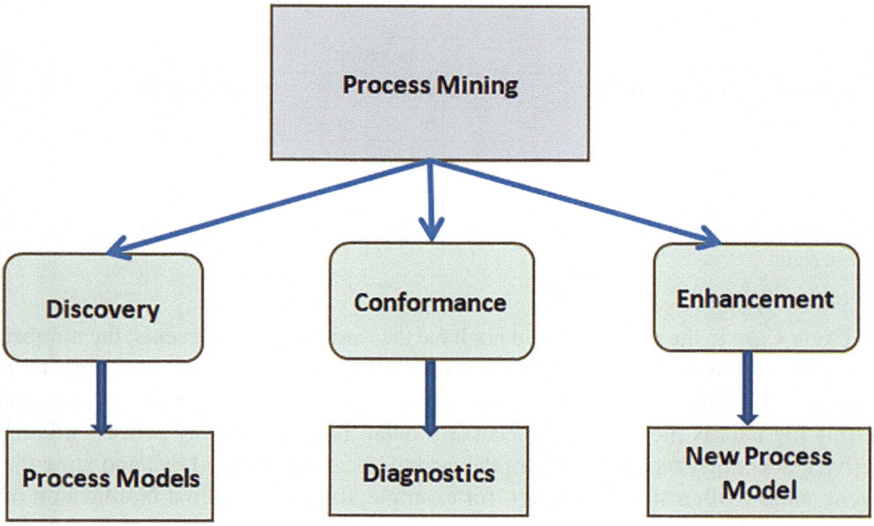

Fig. 2.2 Process mining

Table 2.6 Sample event log

Case ID	Activity ID	Timestamp	Activity	Agent name	Claimant name	Claim amount
1	12341	30-10-2016:8.45	FNOL	Pete	Dale	12345
	12342	5-11-2016:10.45	Assign claim	Pete	Dale	12345
	12343	6-11-2016:10.45	Claim decision	Pete	Dale	12345
	12345	6-11-2016:10.45	Set reserve	Pete	Dale	12345
2	12341	5-11-2016:10.45	FNOL	Steve	Jose	20000
	12342	6-11-2016:10.45	Assign claim	Steve	Jose	20000
	12343	7-11-2016:8.45	Claim decision	Steve	Jose	20000
	12344	9-11-2016:10.45	Payment sent	Steve	Jose	20000
	11346	9-11-2016:11.45	Close claim	Steve	Jose	20000

Table 2.7 Annotated events

Activity	Annotation
First Notification of Loss (FNOL)	a
Assign claim	b
Claim decision	c
Payment sent	d
Set reserve	e
Close claim	f

Every Case in the event log need not have the same number of events; the number of events for each Case can be different. Each Case in the event log contains a sequence of events, this sequence is known as the trace. The information contained in this log data is the minimum necessary detail required for any process mining activity. A simpler representation of the event log can be obtained by annotating the event names with a single alphabet, for example, the letter 'a' First Notification of Loss (FNOL) as shown in Table 2.7. Using this annotation, the events of Case 1 can be shown as <a, b, c, e>, the second Case in the log can be represented as <a, b, c, d, f>.

There are a host of other tools and technologies that overlap with Process Mining. Some of the important ones are Lean Six Sigma, Business Process Management (BPM), Business Process Reengineering (BPR), Data Mining, and Business Intelligence (BI). Business Process Reengineering (BPR) was also a process-centric approach like Process Mining, but BPR aimed at finding a completely new approach rather than proposing improvements in the process. BPR does not analyze data to study the drawbacks in the process, whereas Process Mining is more of a data-driven approach.

Process Mining relies on modelling a process using various notations and formal symbols. There are a variety of widely accepted modelling systems, some of which are discussed here.

Transition System: A Transition system is utilized to describe the behaviour of discrete systems. In its simplest form, it can be defined as a pair $< S, T >$ where S is a set of States and T is a set of Transitions between the states. In a system where a transition from one state to the other occurs on specific actions, then it is represented as a triplet $< S, A, T >$, where A is the action on which the transition occurs. The system will have a few acceptable Start States which can be represented as SStart and a set of final states which can be represented as SEnd, where both SStart and SEnd are subsets of set S. These transition systems have been utilized in the field of computation for quite some time, for example, it was used in operating systems representing a process and its states and in Theoretical Computer Science for Finite State Machines. Essentially any process model that has executable semantics can be modelled as a transition system. Figure 2.3 shows an example of a transition system. It consists of five states S1 through S5, and seven transitions among the states represented with the directed arcs. Each transition occurs on a specific action

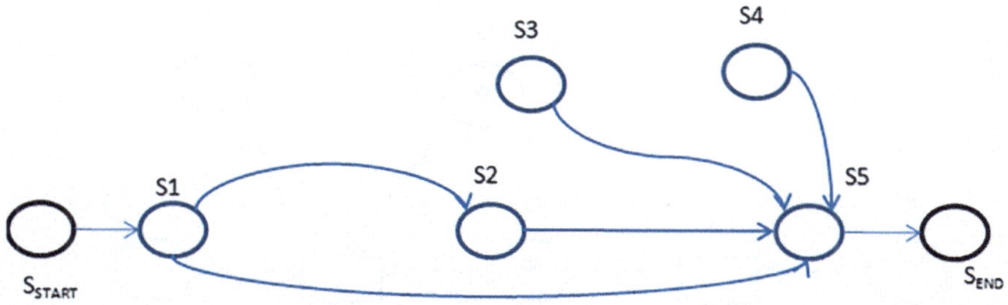

Fig. 2.3 Transition system

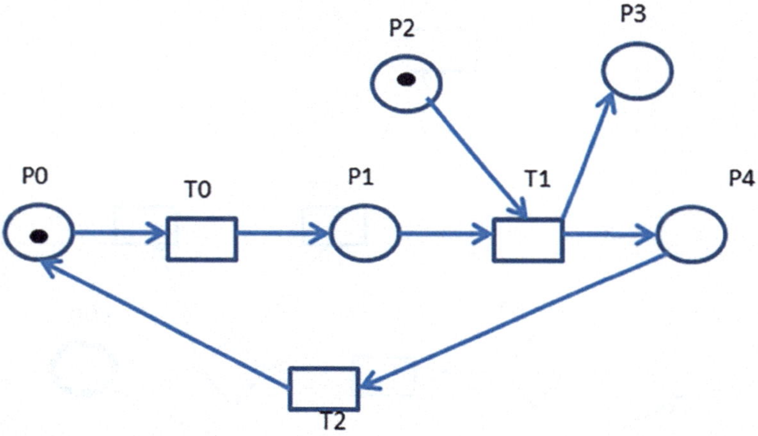

Fig. 2.4 Petri net (Token in P0)

which is labelled over the arcs. The advantage of a transition system is its simplicity and also the interchangeability i.e., a transition system can be easily translated into a UML diagram or a Petri net. The drawback of a transition system is that there can be numerous execution paths from the start state to the end state that can go in parallel, and this concurrency cannot be concisely defined in the transition system.

Petri Net: One of the oldest process models that allow modelling of concurrency is a Petri net. A Petri net consists of places, transitions, and tokens. A firing rule governs the structure of the network. Tokens are distributed over the network, and they determine the state of the network. A simple Petri net is shown in Fig. 2.3, It has five places labeled P0 through P4, three transactions labeled T0 through T2, and two tokens that are present in places P0 and P2. A transaction will fire if it has tokens in its input places, and it will place the token at the output place. Figures 2.4 and 2.5 shows the state of the Petri net before and after the transition T0. When T0 fires, the token is moved from P0 to P1.

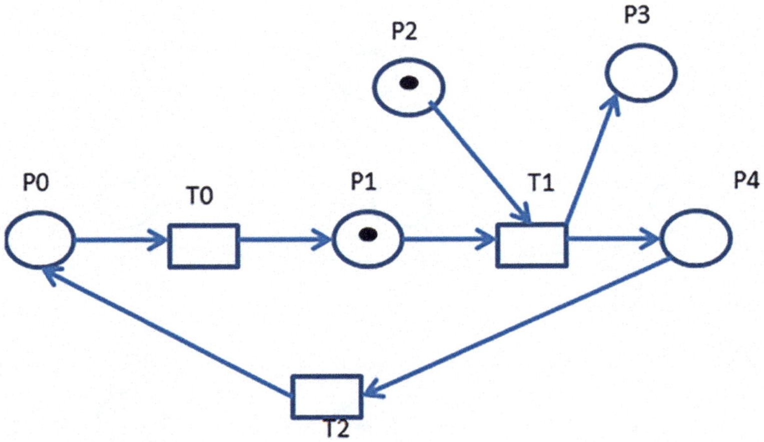

Fig. 2.5 Peti net (T0 fires and token moves to P1)

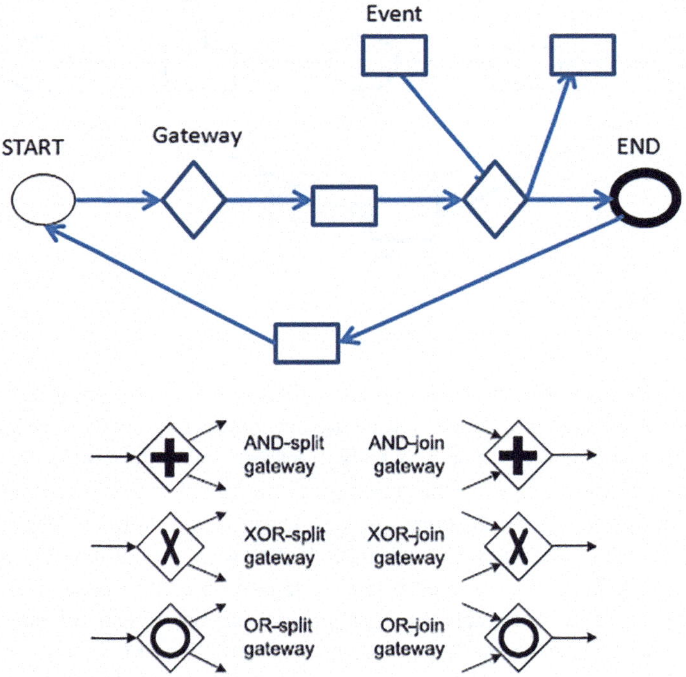

Fig. 2.6 Business process modelling notation

Business Process Modelling Notation(BPMN): BPMN is another widely used system to represent process models. This representation has been supported by many process mining tools. It has a set of notations some of which are depicted in the figure. Activities that are atomic are called tasks and are represented with a rectangle. There are three types of events, the start, intermediate, and end. The start event can have only one outgoing arc and end event can have only one incoming arc, intermediate events have one incoming and one outgoing arc. The routing logic which decides the flow of events is in the gateways. The BPNM diagram shown in Fig. 2.6, has three events and two gateways.

With the exponential growth of event data, Process Mining as a field has come into being thanks to van der Aalst [30–32], who not only pioneered this field but nurtured it with prolific research. Process mining today is applied to various domains such as businesses to healthcare and education. Business Process Mining [33] became popular, as existing tools developed for or Business Activity Monitoring(BAM), Business Operations Monitoring (BOM), and Business Process Intelligence(BPI)were able to give some performance indicators but failed to capture the dependencies and dynamics of the underlying processes.

Business Process Mining (BPM) was able to automatically generate the process models given the event logs. Dustdar et al. [34] have proposed three different perspectives of BPM, the process perspective, the organizational perspective, and the case perspective. The process perspective will focus on the different paths the process can take, its dependencies, and bottlenecks. This can be achieved by building process models and depicting them in the form of Petri Nets. The process perspective is also called the control-flow perspective. The organizational perspective will focus on people involved in the process and how they have been interacting to complete a specific task. Here models such as social networks may come in handy. The third perspective is the case perspective. This is about a particular case, the different events in the case, and the paths it took to complete. This will also help in exploring the alternative paths for a particular case to progress.

There are different approaches and techniques that are used for Business Process Mining, the Transition approach, Clustering techniques, Heuristic approaches, Evolutionary approach, and Declarative Mining techniques. The transition approach was proposed by Aalst et al. in their work [32, 35], where a transition system is first constructed from the log data, and then as a second step, it is modified to avoid any overfitting. Clustering techniques are the approaches [36–38] where the event log is clustered before applying the Business Process Mining. This helps in handling noisy data and also reduces the size of the event logs which in turn will give a smaller and simpler model as compared to the non-clustered log data. Heuristic [39, 40] and Evolutionary approaches [41, 42] are also being widely deployed. Recently, LSTM, Deep Learning, and Transformers [16, 43–45] have also been used in predictive business process monitoring.

Process mining is applied to logs from the Hospital Information System (HIS), to improve the processes in the hospitals [29, 46, 47]. Process mining has also been used to study the models of data for specific diseases such as stroke, cardiac diseases, and cancers [46, 48, 49]. In the Education domain, process mining was deployed to check the efficiency of various ICT tools, to study the student's behaviour, and also to study the usage of the Moodle platform by students [50, 51]. An alternative to event log data which is used as input in process mining is proposed by Goncalves et al. [52] where user stories are used as inputs. Ingvaldsen et al. [53] make use of text-mining techniques to facilitate process mining. The user/citizen stories will also be useful in mining citizen-facing processes. Here citizens may remember only a rough sequence of events and contextual information around that. This is particularly helpful as expecting citizens to provide more information and institutions to make public their internal event logs may not be practical.

Further, even though there is a reference to cross-organizational mining in the literature [32], we came across rather few case studies. Cross-organization mining can occur when a process spans multiple institutions or when multiple institutions follow similar processes. Another view that is less discussed is the information view. Both information view and cross-organizational mining are important for citizen-facing processes as institutions increasingly utilize common information pertaining to citizens, similar processes to interface with citizens, and at times common software packages that have become de facto industry standards.

2.4 Architectural Framework for Managing Citizen Satisfaction

As we have seen in Sect. 2.2, there is a huge body of research that concerns itself with citizen satisfaction. Citizen satisfaction is approached by different scholars from a variety of standpoints. In Fig. 2.7 we represent salient aspects of research outcomes to arrive at an analytical model for citizen satisfaction. The three important enablers of citizen satisfaction are prevailing citizen satisfaction, service quality, and citizen expectations. The service quality in turn comprises technical quality (outcome quality) and functional quality (process quality). The citizen's expectations are driven by perceived value and transparency. Prevailing citizen satisfaction also influences the satisfaction of citizens new to a situation as they start with a positive outlook. An extremely important enabler of citizen satisfaction is flexibility. There are always unforeseen situations where institutions have to demonstrate flexibility. On the flip side, there are bound to be situations where citizens have to be flexible to achieve the outcome they are looking for, without excessive consternation when the processes do not pan out as expected by them. In this work, we propose a new approach where an analytical model guides an architectural framework. This is particularly prudent as most of the processes tend to get automated and come under the purview of software architectural practices, in any case. The architectural approach is extremely useful as it can be done in conjunction with strategic planning exercises.

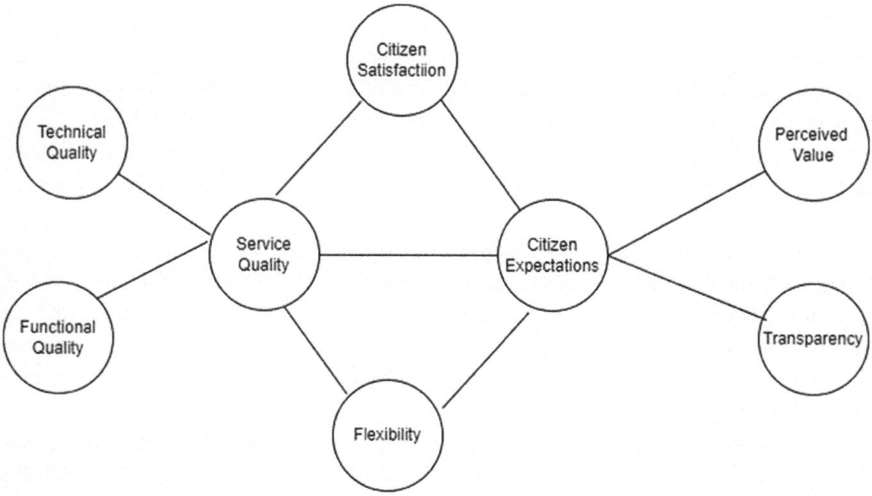

Fig. 2.7 Analytical model for citizen satisfaction

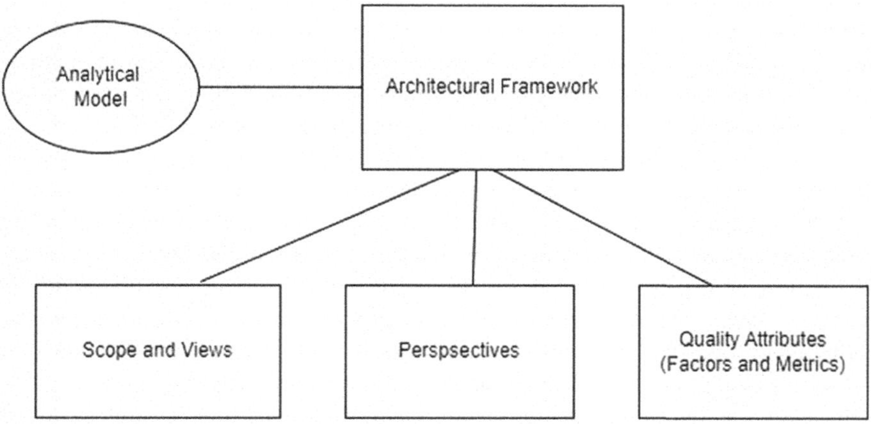

Fig. 2.8 Proposed architectural framework for managing citizen satisfaction

Figure 2.8 illustrates the Proposed Architectural Framework for Managing Citizen Satisfaction. Architectural Framework in turn is detailed in terms of Quality Attributes, Perspectives, and Scopes and Views. Two important goals of any architectural endeavour are addressing stakeholder concerns and quality objectives. Complementing these two are views that are elaborated based on scope. The most important preoccupation with any architectural exercise is catering to stakeholder concerns. Citizens are key stakeholders in any society. Not just satisfying but delighting citizens is a worthwhile goal. Citizen delight will also resonate with politicians, as they can leverage it for electoral success. It will also appeal to the executive management of institutions. In India, it was a long wait for months to get a passport a few years ago. Then the processes were streamlined to such an extent people could get their passports in a matter of weeks. Not all citizens experience many public services directly. But when they hear about good experiences in public service encounters, there is a halo effect, which creates goodwill among citizens It goes without saying that the concerns of other stakeholders such as institutions, employees, regulators, the nation, and other members of society also need to be factored, while delighting citizens.

Citizen satisfaction cannot be analysed in isolation. It depends on an institution's strategic choices. An institution may choose to particularly excel in citizen intimacy, operational effectiveness, or quality of products and services and be on par/lag on the other two. Choosing a strategy also means committing to a set of core values and adopting the right culture in the institutions. Peter Drucker famously said "Culture eats strategy for lunch".

The architectural approach an institution adopts should take into consideration, strategic drivers, efficiency drivers, and value drivers. We used this three-fold approach in our earlier work pertaining to E-Governance Architecture [54]. In this work, strategic drivers are elaborated at the system level, efficiency drivers at the process level, and value drivers are captured along the value dimension. Any institution choosing to excel can learn from how ancient kings managed their kingdoms. They carefully guarded their forts, while keeping a watch on the middle and periphery. In this metaphor, the Strategic Drivers thus map to the fort(system). Efficiency Drivers to the middle (process) and Value Drivers to the periphery (Value). Figure 2.9, elaborates on the architectural framework with stakeholders' perspectives, quality attributes, and indicative actions on how architectural concerns can be addressed.

At the system level, the quality attributes of ethics and social responsibility relate to the core values of an institution. The attributes of communication, consistency, interactivity, transparency, and flexibility are related to work culture and administrative philosophy. Then compliance, security, privacy, and efficiency operate on the strength and robustness of institutional processes and systems. Citizen satisfaction and service quality ideally should be the motto of an institution that is citizen-focused. A discussion on inter-relationships between core values and mottos, strategies, and related is available here [55]. Every institution gets compared with other institutions or itself over time. Hence at every level, we have added relative performance.

The process-level quality attributes are focused on functional quality i.e., how a job is done. Thus, we have from the service quality framework, responsiveness, reliability, assurance, empathy, and tangibles. To this, we add service innovation. There are always opportunities to innovate processes. Service innovation may be also about adding value or additional benefits or customizing or personalizing a service. Additionally, we draw on our earlier research [56] to include Leanness and Agility. Here Leanness is about reducing the scope. It is about striving to have a minimal number of steps. This requires that institutions review the process and analyse what steps are really required in an ongoing basis. In many cases, it may happen that whenever an outlier case happens, the processes are reinforced with additional checks and those checks remain even when they no longer are needed. During process design, certain scenarios were envisaged that do not happen in practice. In such instances,

Scope Level Dimension	Perspective	Quality Attribute
System (Fort) **Strategic Drivers**	Institution	Ethics, Social Responsibility Communication, Consistency, Interactivity, Transparency, Flexibility Compliance, Security, Privacy, Efficiency Citizen Satisfaction, Service Quality Relative Performance
Process **(Middle)** **Efficiency Drivers**	Department	Functional Quality Responsiveness, Assurance, Reliability, Tangibles, Empathy Service Innovation Leanness, Agility Competence Relative Performance
Value **(Periphery)** **Value Drivers**	Citizen	Technical Quality (Satisfaction, Likeability, Outcome Quality) Timely delivery Service Attempts, Hassle-free service, Ping-pong Counter Bribe Paid, Third-party usage Personal factors, Situational factors Positive word of mouth, Halo Effect Contact, Confusion, Fairness Relative Performance

Fig. 2.9 Detailed architectural framework for citizen-facing process mining

we need to take a relook at the processes and make them lean. Agility is to some degree about speed. The essence of agility is finding alternate paths/steps to ensure that the processes conclude within reasonable times. Finally, nothing can replace the competence of service personnel. That is extremely key which many institutions do not pay adequate attention to.

Recruiting the right people and training them to exhibit the right traits is extremely important. Only then service can be provided with requisite Responsiveness, Assurance, Reliability, Empathy, and Tangibles. Even though service quality per se is a cognitive construct, many of the underlying factors are affective and relational. Employees have a role even with Tangibles, in the manner they present and conduct themselves. Generally, there are standard operating procedures for doing a service. Having said that doing any task mechanically is counter-productive. Here the role of supervisors and the quality of oversight becomes extremely important. The next level/dimension is value. This is associated with technical quality or quality of outcome, likeability, and satisfaction. Scoring high on functional quality does not automatically mean high technical quality. To give an example, a restaurant may provide highly courteous and prompt service, but the quality of food may be nothing to write home about. Thus, it is very common to see good restaurants being highly crowded with citizens willing to wait longer for better food. Another important parameter associated with the value dimension is Timely Service (was service received when citizen needed/wanted it?). Did it require multiple attempts? Overall, was the service hassle-free? Many Government Services are complex and without a third party, it is extremely difficult to avail them. In some cases, bribes have to be paid. Thus, on these dimensions, what is the value experienced? Involving third parties, paying bribes, or putting up with inconveniences reduces perceived value and hence citizen satisfaction. Every citizen is different and they all may have varying constraints and challenges. Personal and situational factors impact the value and citizen satisfaction. A citizen may have access barriers. Some of this should be handled by the Process dimension, ensuring that the employees are empathetic.

Another important measure that can frustrate citizens is the Ping Pong Counter. Here the action keeps moving between citizen and institution's employee. The institution's employees are generally adept at passing the buck back to citizens or other departments. At times the corrective action may be to improve communication on what is expected from citizens. In other cases, it is a lack of agility and flexibility in the institution and the process at hand.

When Ping-pong happens between departments it is likely to be because of a lack of clarity in institutional/process design. Clarity can be achieved using the RACI Model. Here those Responsible(R), Accountable(A), Consulted(C), and Informed(I) are clearly identified for every process and case. In a support organization or a bank branch, the manager should generally be informed about all outliers individually and routine cases broadly.

A well-informed and well-prepared citizen may be able to extract higher value in a public service encounter. Any citizen happy with the value he received generally shares with others and this in turn leads to a halo effect and influences citizen satisfaction. This may even influence citizens to assess service quality higher leading to citizen satisfaction becoming antecedent to perceived service quality. Two additional attributes are added to the value dimension i.e., contact and confusion. Contact and Confusion are due to Smith [57]. Contact happens when a transition to the desired outcome (output place) does not happen as it is already/always occupied. Thus, one may never get the service. Confusion prevents a clear distinction between deterministic and indeterministic behaviour. Hence, one does not know if a service request is rejected legitimately or not. Fairness is another interesting measure. When it comes to services, the Fairness concept can be extended to distribution over time, here the resource may be distributed equally, but some get it earlier and others later based on priority. There is an enormous literature [58] on fairness.

The architectural framework should be accompanied by guidelines and should cover critical factors such as communication. Communication is a critical quality attribute that should percolate from institute level to department level to employee to citizen as well in information systems, artifacts, signboards, and everywhere. In the public sphere, citizens and employees co-create a service. Hence citizens need to be well informed about what is needed from their end to participate in the process to avail a service. Also, a citizen should have an idea of how long the process typically takes to fructify. This up-front communication in a manner that is easily understandable by citizens can be considered as an overarching quality attribute that applies to multiple processes. Good communication helps to set the right expectations and is the right thing to do as per disconfirmation theory. However, too much communication should be avoided. Communication which is overly legalistic and technical also does not add value. Communication goes hand-in-hand with transparency. The second important quality attribute is transparency. Transparency is needed before, during, and after any process Firstly, citizens should be made aware of what are the real, actual, and net benefits of availing a service. Even here the right expectations should be set. To give an example, getting a Voter ID in India is an involved process. But to be able to vote, one's name should be in the list of voters of a specific booth. These two processes which should ideally be joined at the hip work in a disjoint manner.

Transparency is about setting the right expectations on how long a process is expected to take and what are the entry criteria for the process as well as what typical intervening steps/sub-processes and how involved they are. When things go wrong, many institutions hide information or become evasive. Many institutions are reluctant to share upfront system-level information related to processes or how processes are operationalized. Citizens are either provided minimal information or given a plethora of information. It is equally important to be transparent when a process leads to long-term engagement. In the case of the life insurance industry, very few companies share that citizens lose a lot of money if they choose to prematurely exit from policies. Unethical companies mislead when they wrongly set citizen expectations.

Yet another significant parameter is flexibility. Even with the best process design and deliberation, there will always be customer situations that need manual overrides or exceptional approvals. Hence, institutions need to have flexibility as a core value. They can also document situations where and how this flexibility is to be used. Process Mining literature makes use of concept drift to indicate that by the time a process is defined to work with a situation or context, the situation or context may change. Redoing the process hurriedly or in an ad hoc manner is not desired. Here flexibility can fill the gap.

Another aspect of transparency is to keep citizens informed whenever there is a state change in the process. Transparency goes hand-in-hand with interactivity. Some institutions do not communicate at all. What is worse some institutions inconsistently communicate. The most important aspect of interactivity is citizens being able to interact with human beings to find out the status and recourse to any action when things go wrong or are delayed. Many institutions seek consent from citizens to terms and conditions at times onerous, at times obvious, and at other times devious. In many such cases, citizens do not have any option but to give their consent. An ethical institution should give citizens real options and not be legalistic about them. A model institution should prepare a set of guidelines that apply to processes, departments, functions, and the institution and move towards a normative state over time.

In current times, a large number of processes are digital and implemented using information systems. Thus, the quality of information systems becomes pertinent. Given in Table 2.8, the criteria for designing a successful information system that

Table 2.8 Designing successful information systems

Model	Factors	Scope
System Quality	accessibility, ease of use, ease of learning, intuitiveness, system reliability, system flexibility, sophistication, and response times.	System
Information Quality	accuracy, relevance, precision, reliability, completeness, usefulness, currency preferred format.	System
Net Benefits	how much the information system adds to the success of the individual, group, organization, industry, or nation	System
Service Quality	support users receive from the service provider	Process
User Satisfaction	how the user feels about the whole experience with the system.	Process
System Use	amount, frequency, nature, extent, and purpose of the use.	Value
Information System Success Model	intention to use, use, user satisfaction, net benefit	Value
Quadratic Usage Framework (QUF)	Dynamics of usage based on personal and environmental factors	Value

makes use of work by DeLone and McLean [59] and work by Mardis [60] on Quadratic Usage Framework. We map the factors to the scope dimensions used above. Here the system use, and usage statistics are important metrics. With systems that allow people to raise complaints, if there is no feedback or communication on change of status, people will stop raising complaints. If the response is rather bureaucratic and does not result in any net benefit, citizens will stop using the systems. If the process of complaining is made very hard, then again people will not complain. Hence low number of complaints should not be considered a sign of a lack of citizen dissatisfaction. Generating confidence in citizens to use the process also is important. Here net benefit accounts for the price and inconvenience citizens have to bear to avail of a service/benefit. In mandatory e-Government contexts, frequent use does not imply system success. Only user satisfaction can fully indicate system success [61]. Table 2.9 illustrates how service quality parameters pertaining to an offline environment can be interpreted for online Information Systems/IT-enabled processes.

Table 2.9 Service quality factors for IT-enabled processes/information systems

Factor	Interpretation in Banking Sector	ITES/IS interpretation
Reliability	Following up on promises in a timely manner; Error-free records and when a problem is reported, bank shows sincere interest in solving it.	Data integrity. Turn-around times as committed. Good citizen support system that is not dogged by bureaucratic approach.
Responsiveness	Employees give prompt service; Employees are always willing to help; Employees will tell exactly when the service will be performed	Good citizen support system that is not dogged by bureaucratic approach. Ability to access using email, and phone without having to go through centralized and unfriendly IVR menus and aloof and inconsistent customer support channels. IT System is responsive, with good help facilities, FAQs, and clear information on whom to contact for what.
Assurance	The behaviour of employees instills confidence; Transaction safety; Consistent Courtesy; Employees knowledgeable to answer questions	Employees are well-trained in IT systems. IT system behaves in such a way that it invokes trust in users. Online resources available contain contextual and domain knowledge that is easy to understand.
Empathy	Understanding specific needs of citizens; Operating hours convenient to citizens; and individual attention.	Well-deliberated user interfaces. Good escalation channels. Single-point interface for a given case till resolution with a backup and point of escalation.
Tangible	Modern-looking equipment; presentable employees; satisfactory physical facilities and amenities	IT Systems are well-designed with a good amount of investment. Support personnel conduct themselves professionally as well in humane manner.

Welch et al. [62] did a study of citizen satisfaction on e-Government Services and how far they have increased the trust of citizens. They found that while citizens are generally satisfied with the electronic provision of information (transparency), there is some dissatisfaction with the transaction (what all they can do in online mode) and interactivity of Web sites. Interactivity is already covered in the framework and transactions are a moving target. Swaid and Wigand [63] consider six service quality dimensions for online services such as website usability, information quality, reliability, responsiveness, assurance, and personalization. In addition to these factors, System Stability is an important concern. Another important factor is Interactive Services (IS) Real-time communication methods, such as online customer support, are available on the website. Providing online forums for citizens to discuss common issues and questions can be facilitated by institutions.

While clarity helps it does not necessarily result in meaningful change. There may continue to be friction. In such cases, it is proposed that a Process Intelligence Group (PIG) be constituted at the institution level. The group comprises Process Champions who work across departments and track processes in an end-to-end manner. Process Champions observe how processes flow through and when they get stuck. They may need to work overtly and covertly. In addition, a Citizen Focus Group (CFG) consisting of citizens and employees from institutions can be formed who have a particular interest in improving the general situation. The PIG and CFG can collaborate with each other. Incentives and rewards can be given to employees and citizens.

In the online world, the click-stream logs can give an idea of how different citizens are accessing IT-enabled processes and their pain points. With online interfaces, it is also easier to determine which fields are seldom filled by citizens, in a form. Such analysis can be used to redesign forms and processes. A citizen may have to repeatedly try an operation, and some way to monitor such cases should be there. With online interfaces, it is also possible to know which citizens coordinate their actions and perpetrate fraud. That is another task PIG and CFG can work on. Analysis of processes can lead to opportunities that Governments need to intervene. For example, standardization of addresses can help streamline a lot of processing, provided legacy issues are addressed thoughtfully and in a planned manner.

The worldview of the Western countries is that society is a collection of individuals. IT systems reflect that worldview. In traditional societies, society is a collection of communities, and communities in turn are a collection of families. Institutions need to go beyond knowing citizens to know their families, localities, and workgroups. Overall, any information system/process should fulfill safety, liveness, and invariance criteria. Here safety ensures that nothing bad happens. Liveness ensures that things do not get stuck. Invariance ensures that there are no leaks in the system due to fraud. Only intended beneficiaries get the benefit, privilege, authorization, and service.

In summary, in this section we have elaborated an architectural framework to manage citizen satisfaction covering stakeholder perspectives, quality attributes, and changes/actions required at multiple levels and interfaces of institutions, covering both manual/offline and automated/online processes/environments.

2.5 Research Model

In this section, we propose a new research model to manage and enhance citizen satisfaction. As a backdrop, in Fig. 2.10, we illustrate the Contemporary Research Model typically used by public institutions. Here analytical model that captures important parameters pertaining to citizen satisfaction, drives the design of citizen satisfaction surveys. Then, the citizen responses may lead to a certain degree of process redesign and institutional redesign. However, a lot more may get done depending on the specific context, when the survey is undertaken and the richness of responses captured.

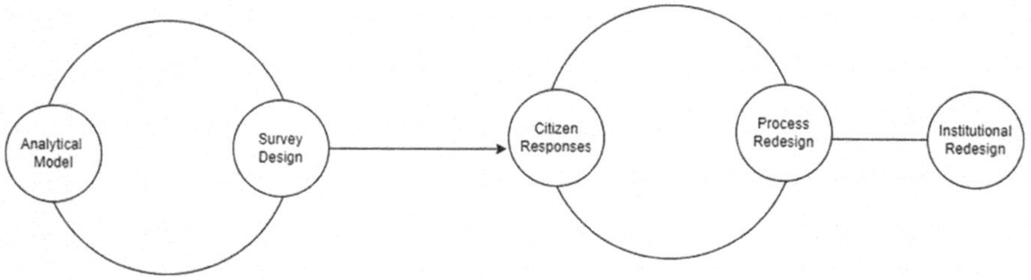

Fig. 2.10 Contemporary research model to manage citizen satisfaction

Some changes institutions may make include changing personnel as well as bridging the gulf between documented process and actual practice.

The research on mutual relationships between service quality, citizen satisfaction, and performance measures has been ongoing since the 1980s. The research model has broadly remained the same where surveys of citizens are the primary input. Generally, Institutions are driven from the top. Thus, the voice of top management and citizens counts whereas rich input from employees themselves does not get pride of place. Further in many institutions employees focus on assigned tasks, jobs, and roles with very limited interaction with other employees and departments on the topic of citizen satisfaction. There indeed are audits as part of the quality process, but many tend to be mechanically and ritualistically done. Hence there is a need for a new research model that engages employees, departments, and institutions in a new way where their human and social impulses are drawn upon. Further, a new research model is necessitated as the contemporary model falls short on multiple counts. Generally, once a process is defined and deployed most institutions seldom bother about how the processes are operating on the ground. They simply deploy a process that seems sensible and leave it at that. They may at most do a periodic survey that gives only broad indicators or a collection of feedback that has comments without context. This may lead to either inaction or a wrong diagnosis with some irrelevant cure and avoidable expenses. They may also plan for limited capacity expansion to serve citizens better and enhance oversight and nothing more. Even here in many institutions, the citizen-facing functions are typically understaffed and barely meet the needs. Even when they have the luxury to do more, they have no inclination or understanding of how they can aspire to do more. They end up doing routine things with hardly any dividends.

Figure 2.11 depicts the new research model we are proposing. The research model is guided by an architectural framework that lays down goals, objectives, and metrics to be pursued at different levels of the institution. The research model makes use of a 3-fold inquiry into reality. There are 3 ways to inquire into reality, according to Reason and Torbert [64]. The First-order inquiry) is research that we do on ourselves. In an institutional setting, it is about assessing what one feels about oneself. Second-order inquiry involves creating communities and groups and research is an outcome of conversations. In Third-order Inquiry), the collective exercise is formalized and institutionalized. According to Joy Sen [65], the Indian tradition adds the

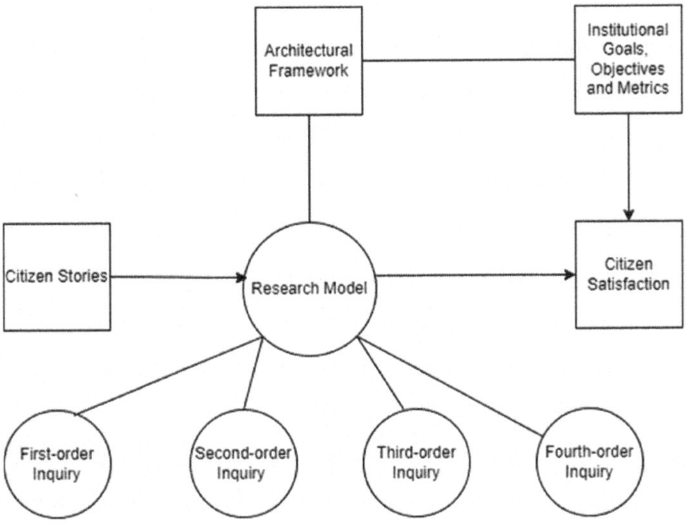

Fig. 2.11 New research model to manage citizen satisfaction

4th dimension to research through "Transcendental Inquiry" wherein researcher and research merge as a whole and there is a heart-to-heart connection between object and subject, thereby expanding the paradigm. Here individuals personify the institution and become the institution. They live the core values, strive to realize the motto, and so on. That is the ultimate pinnacle for the institutions to climb. This we refer to as Fourth-order Inquiry.

The rationale behind this research model is that to achieve citizen delight it is not enough if we do routine things like training support personnel better and overseeing their work more diligently. Rather whole institution needs to be involved. Further, the citizen-satisfaction endeavour calls upon cognitive as well as affective competencies on one hand and system thinking on the other. Fig 2.12 covers a detailed research model with four-fold inquiry. We also include Macro/System Level and Information System Level inquiries for the sake of clarity and completeness.

The research model also takes into consideration that different people seek motivation differently. The impel-compel-propel categorization of people is discussed here [66]. Some thus qualify naturally to be process champions as they are conscientious while serving citizens. Further, today's institutions are social organizations far removed from manufacturing factories where workmen become instruments in the larger scheme of things. In Fig. 2.10, the input to the research model is citizen stories, which indeed are important, but they are only the starting point for analysis. The detailed research model in Table tab10, uncovers stories behind the citizen stories which are symptoms manifested. The diagnosis of stories as well as the prognosis of forthcoming citizen stories needs operationalizing the four-fold inquiries coupled with analysis at the Macro and System Levels. Investing significant time in these inquiries can truly strengthen an institution and impart robustness to all its undertakings.

Nature of Research	Subject	Impulse	Inputs/Artefacts	Expected Outcomes
First-order Inquiry	Employee	What do employees feel when they do their job about themselves	Case dairies, Service call log, Daily Accounts, Success Stories	Common Problems and Issues, Blockers, Ideas how citizen can be helped better
	Citizen	What does a citizen feel during the public service encounter	Anecdotes, Stories, Visitor Logs	Suggestions on how they can be served better. Impediments faced by citizens, Stories where things went well
Second-order Inquiry	Supervisor	Conversations with employees	Daily Accounts, Reports,	Analysis of oversight activities, corrective actions
	Employee Groups Supervisor Groups	Conversations within and across functions	Work-out meetings, Best Practice Sharing, Appreciative Inquiry	Resolving inter-departmental issues, boosting team work. Knowledge Transfer
	Citizen Focus Groups	Conversations with other citizens in/on physical, virtual and social platforms	Experience sharing with others	Enables social learning and expectation formation
Third-order Inquiry	Process-Intelligence Groups	Zeal to improve things.	Process trails, Citizen accounts across departments and timelines. Structured interactions, monitoring, and analysis.	Proactive monitoring Monitoring deadlocks and lost requests. Monitoring delays
Fourth-order Inquiry	Process Champion	Zeal to be a role model. Conscientiousness coupled with professionalism.	Model work in task/job/role	Can guide and inspire others and rest of the organization
Macro/System level	Executives and Managers	Big picture understanding and systems thinking.	System-level conformance, performance, and governance issues and metrics	Steady and continuous improvements, reforms and transformation.
Information System Level	Executives, IT Department	Commitment to develop a system that leads to citizen satisfaction.	Usability System Stability	Continuous Improvements based on feedback from users, click-stream logs, etc.

Fig. 2.12 Detailed research model for citizen satisfaction

In summary, in this section, we have discussed a new research model that involves employees of the institutions and citizens individually, in formal and informal groups, and calls out model employees, citizens, and functions coupled with analysis at the Macro/System Level and Information System level. We are proposing this as a superior alternative to survey-based research models in vogue.

2.6 Analysis of Citizen Stories Using Citizen-Facing Process Mining

In this section, we propose a framework for Citizen-Facing Process Mining and use it in conjunction with research model discussed in Sect. 2.5. In Fig. 2.13, we depict the research and process landscape wherein we are proposing Citizen-Facing Process Mining (Fig. 2.12). As seen in Fig. 2.13, the research into citizen satisfaction kicked off around 1979 with service quality, citizen expectation, and performance measures as key enablers. This research covered offline and manual processes and facilities to start with and continue to date. Over time the processes got automated, starting with back-end services, which broadly cover the period since 1990s. Process Mining as a field took-off around the year 2000 and research has been continuing to date. The focus however is on certain geographies. In the last two decades, with the emergence of the internet and digital inclusion, the processes have become digital and citizens are making online interfaces to access services. We anticipate an emergent need for Citizen-Facing Process mining that covers manual/offline, digital/online, and hybrid processes and expect this to be an active field of research going forward. Even with significant automation and digitization, the focus continues to be on data, and major advances and vocabulary we use pertain to data science. The process mining literature brought the spotlight on process science and started building on well-regarded models such as Petri Nets (Fig. 2.13). According to our observation, the focus of process mining has been the process efficiency and operational improvements without an explicit orientation toward citizen satisfaction. In this work, we attempt to make use of that opportunity by repurposing process mining for citizen satisfaction. Process mining as a field originated at a time when processes were practiced without a clear or documented process model. The other possibility is that the way the process is actually practiced keeps changing based on exigencies. Thus, an important value-add of process mining was to discover processes. With automation, it becomes easier to monitor processes but the citizens and employees of institutions get far removed as they no longer have face-to-face contact.

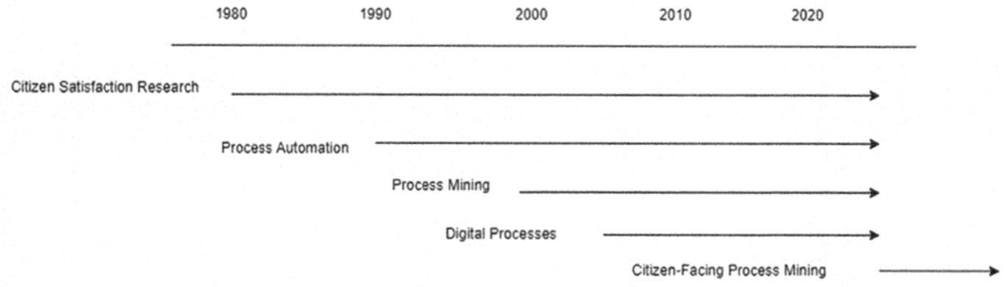

Fig. 2.13 Research and process landscape

Even though process mining is fairly complete in itself with the ability to discover, monitor, and enhance processes, to realize value architectural concerns need to be looked at. At times, it may be the case that an institution's capacity, training, culture, or values are an issue. In such cases, process improvements do not translate into gains. The approach we have taken is to embed processes within the service quality framework and then look at system-level issues and do that exercise with citizen satisfaction as the goal.

Process mining generally makes use of event logs as input and the focus is on operational aspects of institutional processes. The issue with using event logs is many institutions may be oblivious or insensitive to the kind of delays and disconnects the citizens face at their end to meet pre-requisites and co-requisites of a process. Hence, we intend to make use of citizen stories as input with a strong external focus in addition to a reasonable degree of visibility of what plausibly happens within institutions. The processes chosen for study are citizen-facing processes with sufficient visibility to citizens. Here, the citizen stories are analyzed and action plans are associated with employees, departments, focus groups, institutions, or citizens. The overarching assumption is that more than how a process is designed, how a process is practiced on the ground is of paramount importance. While it is important to enhance processes, it is equally important to improvise them. The research model we use should enable both improvisation and redesign of processes.

In institutional process mining, process models are discovered using event logs. Then process conformance studies are done which help in diagnosing problems. This is followed by process enhancement which leads to new process design. In Fig. 2.12, we propose a framework for Citizen-Facing Process Mining. In Citizen-Facing Mining, we use Citizen Stories and discover Citizen-Facing processes. However, the stories may need to be corroborated with event logs of institutional processes. The Citizen-Facing Processes have additional steps that citizens have to take, as there are external dependencies. The Citizen-Facing Process is also the view of the institutional process that is visible to them, Thus, there may be changes to the institutional process, and new rules and guidelines may come into play and thereby cause angst to citizens. On the flip side, institutions also need a model of citizen-facing process

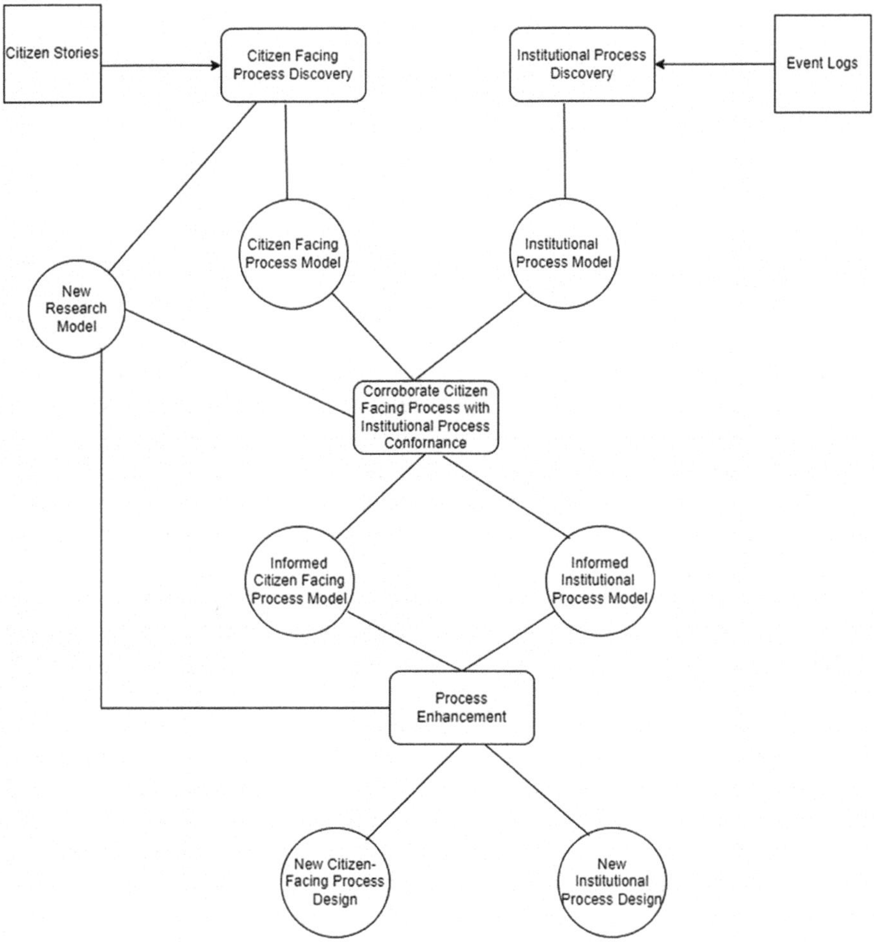

Fig. 2.14 Framework for citizen-facing process mining

that includes steps taken by them outside the purview of the institution to better understand their issues and also to assess how the processes as a whole function (Fig. 2.14).

Institutional Process Steps	Citizen-Facing Process Steps
1. Collect Personal and current Driving License Details and Validate *2. Collect inputs regarding Medical Fitness of the applicant and validate.* *3. Validate that the payment is done.* *4. Validate that the applicant does not have history that disqualifies the applicant* *5. Process the application and generate driving license.* *5.1 Get the required approvals* *5.2 Print the license of media* *5.3 Send the Driving License to addresses*	*1. Keep Personal and Driving License details ready.* *2. Get Medical Fitness Certificate* *2.1 Get Eye Test done* *3. Apply Online* *4. Payment using payment gateway* *5. Await delivery of driving license/Collect Driving License*

Fig. 2.15 Institutional process versus citizen-facing process for renewal of driving license

In Citizen-Facing Process Mining, we start with analyzing citizen stories to discover citizen-facing processes. Then the discovered process is corroborated with the institution process instance captured using event logs. This will help understand the conformance of the institutional process vis-à-vis citizen citizen-facing process. The outcomes of conformance are the Informed Citizen Facing Process and Informed Institutional Process where both citizens and institutions mutually better understand what happens at the institution end and citizen end. The conformance phase is followed by the enhancement phase resulting in new/improved process designs of Citizen-Facing and Institution-Facing Processes. This also may mean revisiting the prerequisites and co-requisites for citizens to participate in the process. All three phases, namely process discovery, process conformance, and process enhancement are done using the new research model. In Fig. 2.15, we explain the difference between the Institutional Process and the Citizen-Facing Process. Here we discuss the renewal of driving license (Fig. 2.15). The citizen who applied was in his 60's. In the case above, citizens may face problems in getting appointments with recognized medical professionals. He may not be aware that an eye test report is mandatory. Some Doctors may certify ignoring eye tests which is not desirable as well. The payment gateway may hang and there could be a variety of problems. All these are best captured using citizen stories. The citizens may not know issues at the institution's end. They may run out of media to print the driving license or if the function is outsourced it may cause delay. Even after all this, the information printed on a driver's license may have inaccuracies, which is hopefully less likely with digital processes.

A good set of questions as used by Rojas et al. [67] where process mining was studied in the context of the healthcare sector were:

- What happened? Identifying the need to discover the process executed and its activities (e.g., what is the typical working day of a surgeon?).
- Why did it happen? Understanding the activities and circumstances characterizing the situation/action (e.g., what caused the long waiting list?).
- What will happen? Identifying the circumstances of when or how a specific activity will take place (e.g., is this patient likely to deviate from the normal treatment plan?). What are the consequences
- What is the best that can happen? Identifying possible steps towards specific improvements (e.g., which check should be undertaken first to reduce flow time?).

Based on this we arrive at the following model to analyse citizen stories.

1. What happened or what typically happens?
2. Why did it happen?
3. What will happen? (Consequences)
4. What can be improved (in this specific case and generally)?

A survey was conducted to collect citizen stories, followed by interactions to get the details. We give below an analysis of citizen stories. Every story is interpreted using the above four questions, analysed, and then the research opportunity is highlighted. Table 2.10 lists the citizen stories along with applicable inquiries from the new research model. Annexure A covers a detailed description and analysis of the stories.

As seen above, the majority of the stories lead us to engage in four-fold research to take a holistic perspective on underlying causes. In certain cases, the issues pertain to information systems and common digital infrastructure used by multiple institutions. This is in contrast to prevailing approaches to address citizen satisfaction which are at times reductionist, at times piecemeal, and at times far removed from ground reality, merely riding on good intent and optimism without adequate validation. Another view of the high-level analysis is given in Table 2.11.

Table 2.10 Citizen stories

Story	Process	What goes wrong?	Research Model Applicable
S001	Raising Police Complaints	Police can register a complaint or record a First Incident Report. Only FIRs are investigated. Police generally refuse to register FIRs and do not act on complaints.	First-order to Fourth-order Inquiries
S002	Right to Information Process in India	Handled bureaucratically and common citizen does not get what he is looking for.	Second-order and Third-order Inquiries
S003	Raising Grievances using the Grievance Portal of Government of India	Officials tend to pass the buck or hastily close the grievance.	First-order to Third-order Inquiries
S004	Interactive Voice Response (IVR) based Citizen Support in Banks	Citizens generally find that they have to devote an enormous time to get to what they want.	Second-order to Fourth-order and Macro Level
S005	Claiming Pension amount while exiting the National Pension Service	The process is overly complex with multiple players involved and automation done without taking into consideration legacy issues	First-order to Fourth-order Inquiries and Information System Level
S006	Withdrawing Employee Provident Fund Amount using Online interface	The process is complex with multiple players involved in the workflow leading to delays and disconnects	Third-order and Fourth-order Inquiries and Information System Level
S007	Driving License Renewal for aged persons	Delay in getting Medical Certification. Payment Hangs and then cannot retry it as it gets locked.	Second-order and Third-order inquiry among citizen groups and employees, Information System Level inquiry and Macro/System Level Inquiry.

(continued)

Table 2.10 (continued)

S008	Getting ATM Card for Joint Accounts and Closing of Joint Accounts	Both need to be present at the same time or authorization should be there. Both should sign on all terms and conditions. Some banks relax the criteria, adding to the confusion,	Second-order Inquiry among supervisors and Institutions. Third-order Inquiry by regulators.
S009	Ordering books online	Exclusive reliance on OTP to verify deliveries	Information System Level/Macro Level Inquiry
S010	Online Furniture Sale	Seller gets cheated by buyer when using UPI interface to receive payment.	Information System Level/Macro Level Inquiry. Third-order Inquiry by Citizen Focus Groups and Regulators. First and Second-order inquiries among citizens.
S011	Using Auto-insurance for repairs	Ethical issues and lack of transparency	First-order inquiry followed by Second-order Inquiry. The third-order inquiry that includes citizen focus groups and Insurance Regulator
S012	Police Verification as part of passport application processing.	In many cases, police instead of visiting citizens ask citizens to visit police stations and do document verification.	Third-order inquiry by institutions and citizen focus groups.
S013	Khata (Property Title) Transfer, Name update	People end up having to pay bribes in certain cases which diminishes the value	Third-order inquiry by regulators and citizen focus groups.
S014	Net banking gets locked due to the wrong password	Here to recover from this one needs an ATM card. In case the person has lost the card or forgot the password he cannot recover access to net banking easily.	Information System/System Level inquiries
S015	Accounts going dormant due to non-use	Here one needs to provide KYC to make the account active by visiting the home branch. This	System/Macro level inquiry on polices

(continued)

Table 2.10 (continued)

		causes lot of inconvenience to citizens.	
S016	Online furniture purchases from branded companies	They started making furniture after order which took a long time and the furniture (TV stand) was not strong enough to support TV mounted on a vertical surface.	Third-order inquiry, Macro/System level Inquiry
S017	Using taxies run by mobile app companies	When there is sexual harassment by drivers, the companies do not take any accountability, Vicarious liability should apply here or heavy penalties should be imposed. They simply suspend the driver and do nothing else.	Third-order Inquiry by Citizen Focus Groups and Regulators. First and Second order inquiries by citizens.
S018	Caste and Income Certificated	By paying bribes one can get certificates that entitle him to benefits and privileges wrongly.	Third-order Inquiry by Citizen Focus Groups and Regulators
S019	FasTag Recharge	FasTag is an RFID-based service at toll gates. The recharge process is inconvenient to citizens as only car owners can make the payment. This may be for security reasons. KYC requirements can make one's account inactive	Information System and System Level Inquiries
S020	Social Media/Platform Use	Governments do not regulate these properly. A lack of due process can put many citizens into trouble.	Third-order Inquiries by Regulators, System/Macro Level Inquiries. First-order Inquires by Citizens
S021	US Visa Processing in India	The rules and regulations are complex and lack of clarity and lack of communication on changes.	Information System/Macro Level Inquiry

Table 2.11 Common citizen satisfaction blockers and related stories

Citizen Satisfaction Blocker	Related Stories	Action needed mainly by
Lack of Strategic Intent to satisfy citizens	S001, S002, S003 and S004	Government, Regulator, and Institutions
Process Design issues	S001, S004, S005, S006, S007, S008, S009, S011, S012, S014, S015, S019 and S021	Institutions/Regulators
Inadequate Oversight (Technical Quality/Outcome poor)	S001, S002, S003, S004	Institutions
Lack of Process Monitoring (Functional Quality Poor)	S001, S004, S005, S006, S007, S008, S012, S017, S019, S021	Institutions
Lack of safeguards in digital public service infrastructure	S005, S009. S010, S019	Regulator
Lack of Sustained Communication to Citizens during the process	S006, S021	Institutions
Bureaucratic Mindset and lack of flexibility	S001, S002, S003, S004, S005, S006, S007, S008	Institutions
Lack of clarity on what is needed on the part of citizens to participate in a process	S005, S006, S008, S021	Institutions
Lack of transparency on how the system/information system works	S019, S021	Institutions
Lack of adequate regulation or oversight	S010, S011, S016, S017, S020	Regulator/Government
Ethical issues	S001, S010, S011, S013 and S018	Institutions, Regulators

2.7 Discussions

The approach to citizen satisfaction we have outlined goes beyond numbers and objectives. It has congruence with the teaching and philosophy of W. Edwards Deming, a household name in quality circles. Deming propounded a fourteen-point management method with steps such as creating constancy of purpose towards improvement, building quality in the process instead of depending on inspections, driving out fear from the workplace, breaking down barriers between departments, eliminating numerical goals and targets instead looking at improving the system as a whole, inculcating pride in workmanship, vigorous program on education and self-improvement and finally making transformation everybody's job [68, 69]. The concepts underlying Deming's method include Visionary Leadership, Internal and External Cooperation, Learning, Process Management, Continuous Improvement, Employee Fulfilment, and Customer Satisfaction. Whereas Deming's work is well-known for the last seven decades, it requires the right mindsets, culture, and values to accept his approach and adopt it beneficially. The benefits of our approach can be enumerated as follows:

1. Most institutions define processes centrally and deploy them in a top-down manner. Then they use surveys to get inputs from citizens and make changes to processes if need be. The institution is seen as a formal organization subject to rules and standard operating procedures. Generally, the employees who are a gold mine of information seldom actively solicited for their views. In the approach we have prescribed, a lot of importance is given to internal conversations. Both formal and informal methods are used to involve employees. By giving importance to individuals and celebrating role models, the institution's potential is tapped a lot more. The institution is as a whole is seen as a social organization. By recommending the formation of citizen focus groups, the citizens become part of the solution.
2. Typically, institutions have action plans that are internally focused and disconnected from what happens at the citizen's end. In this work, we develop a Citizen Facing Process Model and use it in turn to develop a better-informed Institutional Process Model. Thus, any action plan will tend to be Citizen-Centric.
3. The approach proposed by us for analysing Citizen Stories can facilitate the transformation of IT processes, systems, and frameworks. The institutions can take a re-look at pre-requisites and co-requisites for a citizen to participate in a process. They can also communicate better to citizens using IT systems so that citizens are better informed, equipped, and prepared. Communication Strategy can be planned so that the citizens get an appropriate amount of information with clarity and the right frequency.
4. The focus on Citizen Satisfaction that includes process, outcome, transparency, and flexibility, setting and managing expectations can result in far greater value to citizens.

To operationalize the approach described in this work, the following steps are prescribed.

1. Have a clear strategic intent and plan that underscores the importance of citizen satisfaction in the overall scheme of things. Develop a Citizen Satisfaction Charter for the institution.
2. Define roles and responsibilities with clarity as far as tasks and jobs on one hand and responsibility and accountability on the other. Also, ensure the right people are informed and consulted.
3. Set up Citizen Focus Groups and Process Intelligence Groups.
4. Identify Process Champions across the Institution.
5. Establish practices to collect stories, logs, diaries, and as many artifacts either manually or using automation
6. Set up a cadence for meetings when multiple parties are involved and establish regimens for reflection hours for individuals.
7. Reward citizen satisfaction champions and recognize those who embody desired institution values.

8. Conduct inquiries as per the prescribed research model as and when citizen stories are available.
9. Take the outcome of research and work out action plans.

The approach is generic and hence applies to both offline/manual and online/digital systems as they have as much commonality as there are differences.
To move from Citizen Satisfaction to Citizen Delight, there are social and cultural factors to be addressed. To achieve this, institutions, Government, regulators, and society have to move in lock-step. Unless this happens, limitations will remain. One important silver bullet is communication. It is extremely important for institutions to communicate better to citizens so that they are better equipped and prepared to participate in citizen-facing processes. In many cases, the problem is that citizens do not have clear knowledge of what they can expect and what is expected of them. For example, citizens do not have a clear understanding of what inputs are needed from them to close a savings account in a bank. Then they do not have an idea how long the process may take, and how many intervening steps are there. Further, they may not know what are the options for them to receive the proceeds. In some cases, banks may wait a few days before the account is formally closed, as there may be standing instructions on that account. The process for closing joint accounts may require the presence of the second holder or their authorization. Now, without clear expectations, the citizen is bound to be disappointed. Here some banks may pare down some of these steps whereas others may make the experience hard for citizens to go through.

Further, formal institutions do not acknowledge social structure and practices. Families may share their phone numbers i.e.; phone numbers may be in one member's name but used by another. One family member may represent others as he may be more knowledgeable. All these may introduce friction and lead to citizen dissatisfaction. Such cases can be a topic of discussion for Process Intelligence Groups and Citizen Focus Groups to arrive at appropriate recommendations.

Another issue is the lack of citizen choice. All the social platforms, airlines, and mobile app taxis have similar features and services. Thus, citizens do not have a real choice. This holds true for multiple sectors such as Banking, Airlines, Hospitality, etc. In the case of the Government, certain services are mandatory. In cases such as these coalition of citizens needs to work with the Government and concerned Ministries as well as the media to represent them.

Even though the majority of institutions pretty much follow the same script, there are a few institutions that explore innovative ways to engage with citizens. For example, to ensure transaction safety or consent, some use automated recordings of citizens' voices and capture responses to specific queries. They may make routine and repeated transactions easier to do. Such institutions need to be acknowledged and appreciated by citizens lest these will remain isolated bright spots.

Then there are issues that put citizens in trouble. Thankfully though in India, institutions and employees manage fuzziness and lack of clarity well. However, with automated systems, it does not work. In one case, the PIN code of an area was changed, but all artifacts that are used as address proof have the old address.

Such changes are not thought through nor planned well. This points to a lack of adequate brain-storming among institutions that put together IT solutions and among Government agencies about the collateral effect of their actions.

Taking citizen consent for operations is another important concern. When you open a Fixed Deposit, a citizen is asked to give his consent to a huge number of terms and conditions in fine print, by a reputed bank. Many websites ask your consent to use cookies when unless you give your consent you cannot proceed. These have implications on processes, as if a citizen really wants to go through all terms and conditions, it can hardly be done without session timing out. Some of these issues are taken care of if feedback from citizens is proactively sought soon after the end of a digital transaction and processed.

Most citizen satisfaction studies may capture quantitative feedback related to metrics. The institutions seldom look at citizen interface processes and related architectural concerns. Any such interface should have communication to a citizen being triggered whenever an event that causes a change of state happens. Unless acknowledgment is sought, an SMS may not get delivered, an email may end up in the SPAM folder and a letter again may not get delivered. Just sending a message will not guarantee that it reaches the destination, let alone that the citizen has read it and acted on it. Such issues call for honest and detailed discussions among IT personnel, In particular event modelling, an emerging discipline can be useful here.

It is also important to generate confidence in citizens to use processes. Many citizens do not complain because of technical quality considerations whereas others may find the process daunting (say raising a complaint in police station). These lead to issues festering. In certain cases, raising complaints or speaking out can cause reprisals. Thus, many government departments may look good on paper with very few complaints. Dissatisfaction about government officials in democracies will reflect on politicians which leads to another party coming to power, but the systemic issues with governance perpetuate. Our approach with four-fold inquiries should bring to surface all such issues.

In certain cases what is really needed is not process mining, but practice mining. Nowhere the contrast is as stark as in the healthcare sector. When doctors prescribe diagnostic tests just to rule out problems, citizens get exhausted by the very regimen and citizens will not get a cure for the problem they went with either. Research into this field requires experts to engage with citizens and understand their pain. Similar issues dog process checks at the airport. One stray incident of someone carrying an explosive in liquid form may mean the introduction of procedures that inconvenience millions. Many personnel do not appreciate cultural practices and cause trouble to citizens. In cases like this either the fraternity of professionals should be questioned from within and/or outside.

In closing, whereas Citizen Satisfaction is a worthy goal to pursue, some degree of friction, inflexibility, fuzziness, and unpredictability is also important to deter people from committing fraud.

2.8 Conclusions

In this chapter, we have proposed an analytical model for Citizen Satisfaction that draws on insights from extant literature. Then we developed an architectural framework that is informed by the analytical model. The Framework captures architectural concerns such as scope, perspectives, and quality attributes. We have also addressed factors that come into play as processes have become IT-enabled. Further, we have proposed a new holistic research model as an alternative to surveys. Then we have proposed a Citizen-facing Process Mining Framework that takes Citizen Stories as input and then works in conjunction with the existing Institutional Process Mining Framework and enriches it. To collect citizen stories, we made use of public service encounters that were captured by us and our acquaintances. We have analysed selected citizen stories, outlined action plans, and characterized research opportunities. We have outlined the benefits of our approach and the steps required to operationalize it. This chapter emphasizes that the processes are primarily human endeavours and only if people care do things happen. To make improvements leaders and officials, employees, citizens, and institutions need to participate. The onus is also on Information System Architects and Governments to provide the policy framework and shared information infrastructure. Finally, we have discussed social and cultural limitations that need to be navigated to move from citizen satisfaction to citizen delight. Here institutions, regulators, government, and society need to move in lock-step.

2.9 Glossary of Terms

Terms related to Process Mining [33].

- Data Mining: the analysis of (often large) data sets to find unexpected relationships and to summarize the data in ways that provide new insights.
- Business Intelligence (BI): a broad collection of tools and methods that use data to support decision making
- Process Intelligence: a branch of Business Intelligence focusing on Business Process Management.
- Process Mining: techniques, tools, and methods to discover, monitor, and improve real processes (i.e., not assumed processes) by extracting knowledge from event

logs commonly available in today's (information) systems.

- Event: an action recorded in the log, e.g., the start, completion, or cancellation of an activity for a particular process instance.
- Event Log: a collection of events used as input for process mining.
- Process Instance: the entity being handled by the process that is analysed. Examples of process instances are customer orders, insurance claims, loan applications, etc.
- Process Discovery: one of the three basic types of process mining. Based on an event log a process model is learned.
- Conformance Checking: analyzing whether reality, as recorded in a log, conforms to the model and vice versa. Conformance checking is one of the three basic types of process mining.
- Cross-Organizational Process Mining: the application of process mining techniques to event logs originating from different organizations.
- Concept Drift: the phenomenon that processes often change over time. The observed process may gradually (or suddenly) change due to seasonal changes or increased competition, thus complicating analysis. Other Terms
- Lean Six Sigma/Lean Sigma: Lean Six Sigma is a team-focused managerial approach that seeks to improve performance by eliminating resource waste and defects. It combines Six Sigma methods and tools with the lean manufacturing/lean enterprise philosophy.
- Deep Learning: Machine Learning technique that makes use of enormous computing power to infer models from huge amounts of data in place of feature extraction as a prerequisite to learning. Prominent deep learning techniques are Artificial Neural Network, Convolutional Neural Network, and Recurrent Neural Network.
- LSTM (Long short-term memory): A recurrent neural network that can process data sequentially and keep its hidden state through time.
- Transformer: a neural network model that uses attention mechanism to transform one sequence into another one.
- UPI (Unified Payment Interface): An instant payment system that is in use in India where money can be transferred between two bank accounts.

- OTP (One Time Password): Password sent as a part of an SMS message to a phone belonging to a transacting individual. Generally used as a second factor of authentication.

- KYC (Know Your Customer): Regulators mandate that financial institutions validate the basic information about their customers annually such as Identity and Address Proofs.

- FIR (First Information Report): FIR is a document containing the essential details committed by a criminal offence
(Interactive Voice Response): Mechanism used by Customer Support Departments before Citizens can talk to the concerned support person.

- FasTag: an easy-to-use, reloadable tag that is affixed on your vehicle's windscreen. It enables automatic deduction of toll charges, so you can pass through

- IFSC Code: IFSC (Indian Financial System Code) is an 11-digit alphanumeric code that uniquely identifies all banks and their respective branches participating in the online fund transfer process. The IFSC Code is assigned to each bank branch by the Reserve Bank of India (RBI) to monitor the online transfer of money.

- MICR Code: MICR code (Magnetic Ink Character Recognition Code) is a pattern recognition system used mostly by the bank industry to identify the document's originality and to enable the processing and approval of cheques and other papers.

- Aadhaar eSign: An online and paperless approach to electronically signing documents. Aadhaar eSignatures allow citizens with an Aadhaar number and an Aadhaar-linked cell phone number or mail ID to sign documents immediately and from anywhere lawfully.

- PSU (Public Sector Unit): An institution majorly/primarily owned by the Government that operates in sectors where private players also may operate.

- NPCI (National Payment Corporation of India): A PSU that manages digital payments by acting as the intermediary between banks.

- RTI (Right to Information): This is an Act of the Government of India that enables an Indian Citizen to request information from the Government in the public interest.

- PMO (Prime Minister's Office): PMO in India includes the anti-corruption unit and the public wing dealing with grievances.
Terms (From this chapter)
- Institutional Process: The view of the process as seen from within the institution

- Citizen-Facing Process: The view of the process as seen by the citizen
- Citizen-Facing Process Mining: Applying process mining to Citizen Facing Process in conjunction with the corresponding Institutional Process
- New Research Model: Research Model that encompasses four-fold inquiries and inquiry at Macro/System Level as well as Information System Level
- Leanness of Process: Having minimal or only the required number of steps that are actually needed by weeding out steps that are no longer needed from time to time.
- Agility of Process: Ability to find more than one path so that the processes progress and conclude in a timely manner.
- Fairness: Every Citizen gets serviced within a reasonable time and does not suffer unduly delay. This applies to other quality of service as well.
- Contact: This happens when a transition to the desired outcome (output place) does not at all happen as it is already/always occupied by/due to somebody/some reason or other.
- Confusion: Prevents a clear distinction between deterministic and indeterministic behaviour. Hence, one does not know if a service request is legitimately rejected.

Acknowledgements We sincerely thank the editors for giving us the opportunity to contribute to this chapter and the anonymous reviewers for their comments that helped us to improve this chapter.

Appendix: Analysis of Citizen Stories

See Figs. 2.16, 2.17, 2.18, 2.19, 2.20, 2.21, 2.22, 2.23, 2.24, 2.25 and 2.26.

Story No	S001
Process	Raising Police Complaints
What happened?	Citizens approach the police station when they lose something. Police register a complaint and not a FIR (First Information report). Only FIRs set criminal processes in motion. Even to raise complaints bribes are asked. Once a complaint is filed there may not be any action. In the case of accidents, the affected party is asked to keep the vehicles with the police for long periods.
Why did it happen?	There are two categories that enable police to evade responsibility.
What will happen?	Citizens will avoid complaining which can result in bigger crimes and accidents.
What can be improved?	Citizen visits or interactions with police stations can be tracked and data collected. A third-party agency can be used here or Process Intelligence Group can collect the information. The dichotomy between complaints and FIRs can be done away. Accident cases can be handled better by working with insurance companies.
Analysis	At the system level, ethics and compliance issues. At the process level, lack of empathy. Tangibles also is an issue, as many police stations do not inspire the trust of citizens, but things are improving. Value (technical quality) to citizens diminishes due to low/no outcomes and bribes. In many cases, third parties are needed to get proper service.
Research Opportunity	A Third-order inquiry is needed to understand systemic issues. First and Second-order inquiry is needed to instill ethical behavior. A Fourth-order inquiry to showcase model behavior among personnel can be resorted to change mindsets.

Fig. 2.16 Analysis of citizen stories: story No. S001

Story No	S002
Process	Right to infomation Process (RTI)
What happened?	An RTI request was raised to know what should be the retirement age for Teaching Professionals with the Kamataka Govemment. The process itself was manual and happened as per timelines. The request was rejected saying that the official did not understand what was sought.
Why did it happen?	The officials have a bureaucratic mindset and do minimal actions.
What will happen?	Citizens are dissuaded from filing RTIs.
What can be improved?	There can be greater oversight on how cases are handled. The process can be enhanced so that it can lead to citizen delight.
Analysis	At the system level, things are in place. Scores well on responsiveness. At process level assurance is weak. Here even though Function Quality of service is good, the Technical Quality is bad which affects value. The system should be online/
Research Opportunity	Greater opportunity for Second-order Inquiry and Third-order Inquiry. The inquiry is also required at the citizen's end.

Fig. 2.17 Analysis of citizen stories: story No. S002

Story No	S003
Process	Grievance Portal
What happened?	A grievance was raised regarding the indiscriminate towing of vehicles with the State Government and the Ministry of Road Transport and Highways. After waiting for 14 days, the grievance was closed. Then the grievance was raised with PMO which got addressed. Attempt to raise a grievance with the state government did not succeed. A suggestion was raised with PMO which was closed within a day, saying the suggestion was noted.
Why did it happen?	The officials have a bureaucratic mindset and do minimal actions. Not all ministries and arms of government are equally committed to serving citizens well.There needs to be an interface and adequate capacity to handle suggestions as well in light of participative government. There needs to be officials and government that genuinely value citizen input.
What will happen?	Citizens are dissuaded to raise grievances. The number of grievances to PMO will increase. Citizens get disconnected from government and loss of opportunity to do ground-up reforms.
What can be improved?	There can be greater oversight on how grievances are handled coupled with longitudinal studies.
Analysis	System-level things are in place at the Central Government level, but not at state level. At the process level scores high on responsiveness. At the process level lack of empathy. Value is impacted due to low technical quality. Officials may point to other departments leading to high ping-pong counter.
Research Opportunity	Opportunity for First-order inquiry regarding empathy, followed bysecond and third-order inquiry regarding value.

Fig. 2.18 Analysis of citizen stories: story No. S003

Story No	S004
Process	IVR based Citizen Support in Banks
What happened?	It takes long to reach the option you are looking for. Very hard to get to talk to someone. Unable to get what you want. You get logged out after 2-3 tries. Not able to handle uncommon questions. The wide difference between knowledge and competency levels of support persons in providing resolution.
Why did it happen?	As almost all banks follow the same process, these processes are adopted mechanically.
What will happen?	The citizens will avoid using this channel unless they have no choice. They do not have a choice to migrate to other banks, as all follow similar approaches. The investment made by banks will not give them good return as far as citizen satisfaction is concerned.
What can be improved?	Citizens prefer to talk to a support person than going through IVR. Fewer but more knowledgeable and smarter support persons may be a better option. Provide ability for citizens to call their home branches and get better support.
Analysis	At the system level, a greater focus on citizen satisfaction is needed. Citizen support tends to be seen as a cost that needs to be minimized using maximal automation. At the process level, a major concern is responsiveness (takes too long). Secondary concerns are assurance (knowledge level) and empathy (ability to understand what citizen wants). Lack of flexibility with pre-configured options is another issue.
Research Opportunity	Generally, support persons do their job sincerely, yet fail to satisfy citizens. Second-order inquiry can improve matters within the system, with more conversations. Here a third order inquiry is required to look at the overall process. The macro-level analysis is required to see how the channel can be radically changed or supplemented with branch-level support and humans in the loop. An alternate recourse should be provided to someone who fails to get what he wants using the IVR interface. Encourage 4th order inquiry and celebrate role model employees who work with citizens and fulfill their needs.

Fig. 2.19 Analysis of citizen stories: story No. S004

Story No	S005
Process	Claiming Pension amount while exiting National Pension Service
What happened?	The Pension Account was seeded by capturing legacy information about the bank accounts of citizens which had only MICR code but not IFSC code. This worked fine when used for a depository account with the same bank. Attempting to change bank account details requires Aadhaar eSign which requires the name in bank account to match the name registered with Aadhaar. The name in the bank account abbreviated the middle name and eSign failed.
Why did it happen?	The legacy issues were not considered when Aadhaar eSign was used as the sole mechanism. Manual overrides were not considered where approval can be sought with authorized personnel, as part of design.
What will happen?	In this case, the citizen had to change the name in the Pension Account after a protracted and long delay, and then only the claim could be made. This took around 6 months. Even using a manual process did not help as the final check was using Aadhaar eSign.
What can be improved?	Allowing Aliases when middle names are abbreviated. Allowing validation using PAN cards or manual overrides with authorization. When a new account is created take extra care that all related information is completely seeded. The number of institutions involved is too many. Compared to this Life Insurance Corporation (LIC) handles pensions particularly well, The pension dispersal can be outsourced to LIC.
Analysis	The process design led to deadlock. The process of claiming itself was lean with minimal steps. Agility was low, as there were no alternative routes to recover from deadlock. At the system level, there was a lack of transparency with multiple stakeholders involved and there was no single support channel. The ping-pong counter was high as it was not clear whom to approach for what. There was a lack of assurance and empathy throughout. The emails were mechanically answered. Technical Quality was low due to many service attempts and hassles involved.
Research Opportunity	A greater amount of First-order inquiry on such outliers is needed. Second-order inquiry in the concerned department with planned conversations is required. Any suggestions then can be forwarded to the concerned department involved in IT-enabled process designs as well as senior management. At the system level citizen satisfaction can be tracked using surveys. There should be proactive measures to ensure citizens get money within a certain time, even if it means going out of the way. Encourage Fourth-order inquiry and celebrate role model employees who work with citizens and fulfill their needs. There should also be Information System Level Inquiry.

Fig. 2.20 Analysis of citizen stories: story No. S005

Story No	S006
Process	Withdrawing Employee Provident Fund Amount using Online interface
What happened?	Certain operations were to be done at the employer's end which were not done. In this case, a citizen was transferred between two legal entities belonging to the same management. Appropriate updates were not done by the employer. Then some more time was lost when a particular form was to be uploaded by the citizen. There was inconsistent communication/behavior. In the first case, there was an email alerting the citizen that the claim did not go through. In the second case, there was no email when claim process got stalled, and only after citizen logged in, he could claim the amount.
Why did it happen?	Lack of processes at the employer end when a number of employees got transferred from one entity to another, to ensure that appropriate action was taken to ease the claim process. Every leaving employee had to personally get it done. The issue with the information system and the unreliability of email as an alert mechanism. Acknowledgment should have been sought from citizens when such critical emails were sent. A letter should have been mailed to the citizen.
What will happen?	Citizens who have a choice may not go for Provident Fund, considering Functional Quality in mind to some degree. However, the process was executed in such a way that there was no loss to the citizen. Technical Quality i.e., value to the citizen by investing in Provident Fund continues to be valid.
What can be improved?	More knowledge can be given to citizens so that they can co-create this service with greater preparation. More humans in the loop are required to support citizens as the process is complex
Analysis	Functional Quality can be improved. There was some ping-pong between the service provider and the citizen. Technical Quality was good. There was no issue with Assurance, Empathy, or Reliability. Related to value dimension timely delivery should be tracked. A citizen may have an expectation to get the money within 8 Weeks.
Research Opportunity	A good degree of Third-order Inquiry is required here is to understand how these processes pan out in an end-to-end manner. One should also look for lost processes, where the ball is in the citizen's court and no response is there from him. There should be proactive measures to ensure citizens get money within a certain time, even if it means going out of the way. Encourage fourth-order inquiry and celebrate role-model employees. There should be Information System level inquiry to ensure consistent and event-driven communication to users.

Fig. 2.21 Analysis of citizen stories: story No. S001

Story No	S007
Process	Driving License Renewal due to age
What happened?	A lot of time was spent to get the medical test, which was allowed only in Government Hospitals. This required an eye test and reports from that. The payment gateway hung and it was tagged with the Driving license number and the operation could not be retried. Finally, it required a personal visit to do the renewal with an archaic procedure.
Why did it happen?	With manual processes such operations are done using third parties who in turn may give kickbacks to the employees. Hence, the motivation to radically improve things is low. The second reason is many systems are deployed without adequate acceptance testing and stress testing.
What will happen?	Citizens may continue to engage third parties till online systems are working faultlessly.
What can be improved?	Procedures can be changed to allow doctors in private hospitals and clinics to certify, with an appropriate process.
Analysis	Analysis can be done related to the value dimension regarding third-party usage, hassle-free service, and timely delivery as well as bribes.
Research Opportunity	Information System Level Inquiry to handle IT issues. Third-order inquiry to review medical test requirements as well as age. Second-order and Third-order inquiry among citizen groups,

Fig. 2.22 Analysis of citizen stories: story No. S007

Story No	S008
Process	Getting ATM Card for Joint Accounts and Closing of Joint Accounts
What happened?	In case of one bank, the ATM card was issued to the first account holder. In the case of another bank, the signatures and authorization from both holders were insisted upon, which resulted in delays. Closing of the joint account required authorization from the second holder. In one case the application was rejected as joint holder signed only in one place when signatures were required in two places.
Why did it happen?	Lack of clear understanding among citizens about the joint operation of accounts. Poor design of forms and administrative rigidity. Lack of process to involve supervisors to get their approval in case of genuine cases.
What will happen?	Citizens may choose to move to banks with relatively lax procedures.
What can be improved?	Certain processes can be tailored or simplified depending on how well the bank knows citizens and their families. It is also possible to call the joint holder and get confirmation post-facto. Even email communication and letters can be sent to joint holders. It is possible to delight citizens without leading to security concerns.
Analysis	Processes can be reviewed for both leanness and agility.
Research Opportunity	This requires Second-order inquiry where employees and supervisors can have conversations about citizens' pain points. A visitor log can be introduced in banks to capture visitor feedback. This is also a good case for cross-institutional process mining, which is conversations among institutions using Second-order or Third-order Inquiries.

Fig. 2.23 Analysis of citizen stories: story No. S008

Story No	S009
Process	Ordering Books Online
What happened?	A citizen ordered a book that was to be delivered at his place of work. The citizen asked for the book to be delivered at the gate to a security person. The delivery executive asked for OTP from a citizen and claimed he delivered. But when the citizen went to the gate after a day or two no book appears to have been delivered.
Why did it happen?	Lack of care by the citizen where he should have not given OTP directly to the delivery person.
What will happen?	Citizens will choose not to get the delivery at that location or choose to go to another platform.
What can be improved?	There was no mechanism to reach out to the concerned delivery person and understand what really happened. Without such clarity and diagnosis, any corrective action cannot be formulated.
Analysis	This is a reliability issue at the Process level. It is possibly one-off issue.
Research Opportunity	Information system-level inquiry can be done Over-reliance on OTP as a validation mechanism can prove to be a single-point failure. Additional mechanisms may be needed such as signatures on touchpad's even with OTP. This is a good case for Third-order inquiry as well as at the system level with analytics on non-deliveries.

Fig. 2.24 Analysis of citizen stories: story No. S009

Story No	S010
Process	Online Furniture Sale
What happened?	A citizen was keen to sell her furniture. An online platform connected her with a buyer. The buyer asked her account information to transfer money and instead of paying money, and managed to withdraw money,
Why did it happen?	Lack of knowledge among citizens. Lack of reliable framework to do safe online payments with the ability for recovery actions.
What will happen?	Citizens will refrain from using such platforms.
What can be improved?	Even though UPI is very easy to use, it is also very easy to perpetrate fraud. There can be a cooling period before a person who gets the money can withdraw and use it (24 hours to 48 hours depending on the amount.
Analysis	Security level issue at the system level. Functional Quality is low as there should be mechanisms to look for frauds and online platforms should assume some degree of liability or have insurance mechanisms.
Research Opportunity	First-order inquiry is extremely important here. A citizen should document exactly what happened and how they got cheated. This then should be coupled with information system inquiry to design better systems. Second-order inquiry among citizens using conversations and formal citizen institutions also can help in advocacy for more reliable systems. System-levelInquiry for NPCI (National Payment Corporation of India) as well as online platforms for providing safeguards to their processes and systems.

Fig. 2.25 Analysis of citizen stories: story No. S010

Story No	S011
Process	Repairing Cars using Auto-insurance
What happened?	A citizen's car had a relatively minor accident, where the bumper of the car was damaged. The citizen expected the repair to cost less than Rs 10,000. The service person. put multiple items in the repair scope and got the insurance company to approve the expenses. The citizen was not informed how much he had to pay till the car was ready for delivery.
Why did it happen?	The dealer wanted to inflate the claim as much as possible to enhance their revenue without bothering about the burden on citizens. They did not have a clear-cut process to keep citizens informed in every case. There was a lack of oversight. There was a lack of due diligence on behalf of the insurance company.
What will happen?	Citizens will be very wary of using auto insurance
What can be improved?	There are gaps in the process where getting formal approval from the insurance company is required but no such approval from the citizen for his part of the payment was sought. Lack of process with split invoice
Analysis	Responsiveness was good. There seemed to be an issue with assurance as the employee went ahead with the repair without informing the citizen. Understanding of personal and situational factors was low. Even when escalated the management did not step in to address the concern of citizens and asked for full payment, even as a gesture or token, Overall ethics seemed to be an issue across the sector.
Research Opportunity	Third-order inquiry involving groups of citizens and insurance companies. First-order inquiry where they maintain logs/dairies in case they are being forced to act unethically.

Fig. 2.26 Analysis of citizen stories: story No. S011

References

1. Parasuraman, A., Zeithaml, V. A., & Berry, L. L. (1985). A conceptual model of service quality and its implications for future research. *Journal of Marketing, 49*(4), 41–50. https://doi.org/10.2307/1251430
2. Parasuraman, A., Zeithaml, V. A., & Berry, L. (1988). Servqual: A multiple-item scale for measuring consumer perceptions of service quality *64*(1), 12–40 https://doi.org/psycnet.apa.org/record/1989-10632-001
3. Parasuraman, A., Berry, L. L., & Zeithaml, V. A. (1991). Understanding customer expectations of service. *MIT Sloan Management Review*. https://doi.org/sloanreview.mit.edu/x/3234
4. Parasuraman, A., Zeithaml, V. A., & Malhotra, A. (2005). ES-QUAL: A multiple-item scale for assessing electronic service quality. *Journal of Service Research, 7*(3), 213–233. https://doi.org/10.1177/1094670504271156
5. Oliver, R. L. (1980). A cognitive model of the antecedents and consequences of satisfaction decisions. *Journal of Marketing Research, 17*(4), 460–469. https://doi.org/10.1177/002224378001700405
6. Oliver, R. L. (2014). *Satisfaction: A behavioral perspective on the consumer: A behavioral perspective on the consumer*. Routledge. https://doi.org/10.4324/9781315700892
7. Andrews, R., Boyne, G. A., Meier, K. J., O'Toole, L. J., Jr., & Walker, R. M. (2005). Representative bureaucracy, organizational strategy, and public service performance: An empirical analysis of English local government. *Journal of Public Administration Research and Theory, 15*(4), 489–504. https://doi.org/10.1093/jopart/mui032
8. Andrews, R., Boyne, G. A., & Walker, R. M. (2006). Subjective and objective measures of organizational performance: An empirical exploration. In *Public service performance: Perspectives on measurement and management* (pp. 14–34). https://doi.org/10.1017/CBO9780511488511.002
9. Andrews, R., & Boyne, G. (2010). Better public services: The moral purpose of public management research? *Discussants: Christopher hood, owen hughes and don kettl*. https://doi.org/10.1080/14719030903286656
10. Wichowsky, A., & Moynihan, D. P. (2008). Measuring how administration shapes citizenship: A policy feedback perspective on performance management. *Public Administration Review, 68*(5), 908–920. https://doi.org/10.1111/j.1540-6210.2008.00931.x
11. Reddick, C. G., Perlman, B. J., & Demir, T. (2022). Citizen satisfaction with public service delivery: A test of alternative theories. *Canadian Public Administration, 65*(2), 352–371. https://doi.org/10.1111/capa.12464
12. Kelly, J. M. (2003). Citizen satisfaction and administrative performance measures: Is there really a link? *Urban Affairs Review, 38*(6), 855–866. https://doi.org/10.1177/10780874030380060
13. Mengi, P. (2009). Customer satisfaction with service quality: An empirical study of public and private sector banks. *IUP Journal of Management Research, 8*(9). https://doi.org/www.iupindia.in/909/IJMR_Customer_Satisfaction_7.html
14. Im, T., & Lee, S. J. (2012). Does management performance impact citizen satisfaction? *The American Review of Public Administration, 42*(4), 419–436. https://doi.org/10.1177/0275074011408589
15. Walle, S. (2018). Explaining citizen satisfaction and dissatisfaction with public services. In *The Palgrave handbook of public administration and management in Europe* (pp. 227–241) . https://doi.org/10.1057/978-1-137-55269-3_11
16. Bradford, B., Jackson, J., & Stanko, E. A. (2009). Contact and confidence: Revisiting the impact of public encounters with the police. *Policing & Society, 19*(1), 20–46. https://doi.org/10.1080/10439460802457594
17. Cohen, J. E. (2006). Citizen satisfaction with contacting government on the internet. *Information Polity, 11*(1), 51–65. https://doi.org/10.3233/IP-2006-0083

18. Chatterjee, R., & Suy, R. (2019). An overview of citizen satisfaction with public service: Based on the model of expectancy disconfirmation. *Open Journal of Social Sciences, 7*(4), 243–258. https://doi.org/10.4236/jss.2019.74019
19. Mohammad, A. A. S., & Alhamadani, S. Y. M. (2011). Service quality perspectives and customer satisfaction in commercial banks working in Jordan. *Middle Eastern Finance and Economics, 14*(1), 60–72. https://doi.org/www.iosrjournals.org/iosr-jbm/papers/Vol7-issue2/C0722633.pdf
20. Mistry, D. S. H. (2013). Measuring customer satisfaction in banking sector: With special reference to banks of Surat City. *Asia Pacific Journal of Marketing & Management Review_ISSN, 2319*, 2836. https://doi.org/docplayer.net/21436990-Measuring-customer-satisfaction-in-banking-sector-/with-special-reference-to-banks-of-surat-city.html
21. Singh, J., & Kaur, G. (2011). Customer satisfaction and universal banks: An empirical study. *International Journal of Commerce and Management, 21*(4), 327–348. https://doi.org/10.1108/10569211111189356
22. Fida, B. A., Ahmed, U., Al-Balushi, Y., & Singh, D. (2020). Impact of service quality on customer loyalty and customer satisfaction in Islamic banks in the sultanate of Oman. *Sage Open, 10*(2), 2158244020919517. https://doi.org/10.1177/2158244020919517
23. Jamal, A., & Naser, K. (2002). Customer satisfaction and retail banking: An assessment of some of the key antecedents of customer satisfaction in retail banking. *International Journal of Bank Marketing, 20*(4), 146–160. https://doi.org/10.1108/02652320210432936
24. Grönroos, C. (1984). A service quality model and its marketing implications. *European Journal of Marketing, 18*(4), 36–44. https://doi.org/10.1108/EUM0000000004784
25. Engdaw, B. D. (2020). The impact of quality public service delivery on customer satisfaction in Bahir Dar city administration: The case of ginbot 20 sub-city. *International Journal of Public Administration, 43*(7), 644–654. https://doi.org/10.1080/01900692.2019.1644520
26. Lamsal, B. P., & Gupta, A. K. (2022). Citizen satisfaction with public service: What factors drive? *Policy & Governance Review, 6*(1), 78–89. https://doi.org/10.30589/pgr.v6i1.470
27. James, O., & Wilson, D. (2010). Introduction: Evidence from the comparison of public service performance. *Evaluation, 16*(1), 5–12. https://doi.org/10.1177/1356389009350022
28. Yang, Y. (2018). Is transparency a double-edged sword in citizen satisfaction with public service? Evidence from china's public healthcare. *Journal of Service Theory and Practice, 28*(4), 484–506. https://doi.org/10.1108/JSTP-06-2017-0102
29. Royne Stafford, M., Stafford, T. F., & Wells, B. P. (1998). Determinants of service quality and satisfaction in the auto casualty claims process. *Journal of Services Marketing, 12*(6), 426–440.
30. Van Der Aalst, W., & Aalst, W. (2016). *Data science in action*. Springer. https://doi.org/10.1007/978-3-662-49851-4_1
31. Aalst, W. M. (2014). Process mining in the large: A tutorial. In *Business Intelligence: Third European Summer School, eBISS 2013, Dagstuhl Castle, Germany, July 7-12, 2013, Tutorial Lectures* (Vol. 3, pp. 33–76). https://doi.org/10.1007/978-3-319-05461-2_2
32. Van Der Aalst, W., Adriansyah, A., De Medeiros, A.K.A., Arcieri, F., Baier, T., Blickle, T., Bose, J.C., Van Den Brand, P., Brandtjen, R., & Buijs, J. (2012). Process mining manifesto. In *Business Process Management Workshops: BPM 2011 International Workshops, Clermont-Ferrand, France, August 29, 2011, Revised Selected Papers, Part I 9* (pp. 169–194). Springer. https://doi.org/10.1007/978-3-642-28108-2_19
33. Van Der Aalst, W. M., Reijers, H. A., Weijters, A. J., Dongen, B. F., De Medeiros, A. A., Song, M., & Verbeek, H. (2007). Business process mining: An industrial application. *Information Systems, 32*(5), 713–732. https://doi.org/10.1016/j.is.2006.05.003
34. Dustdar, S., Hoffmann, T., & Aalst, W. (2005). Mining of ad-hoc business processes with teamlog. *Data & Knowledge Engineering, 55*(2), 129–158. https://doi.org/10.1016/j.datak.2005.02.002
35. Rubin, V., Dongen, B., Kindler, E., & Günther, C. (2006). Process mining: A two-step approach using transition systems and regions. BPM Center Report BPM-06-30, BPM Center, Technical Report

36. Aalst, W. M., Rubin, V., Verbeek, H., Dongen, B. F., Kindler, E., & Günther, C. W. (2010). Process mining: A two-step approach to balance between underfitting and overfitting. *Software & Systems Modeling, 9*, 87–111. https://doi.org/10.1007/s10270-008-0106-z
37. Ferreira, D. R. (2009). Applied sequence clustering techniques for process mining. In: *Handbook of research on business process modeling* (pp. 481–502). IGI Global. https://doi.org/10.4018/978-1-60566-288-6.ch022
38. Zandkarimi, F., Rehse, J.-R., Soudmand, P., Hoehle, H. (2020). A generic framework for trace clustering in process mining. In *2020 2nd International Conference on Process Mining* (ICPM) (pp. 177–184). https://doi.org/10.6084/m9.figshare.12607742.v2. IEEE
39. Aalst, W., Weijters, T., & Maruster, L. (2004). Workflow mining: Discovering process models from event logs. *IEEE Transactions on Knowledge and Data Engineering, 16*(9), 1128–1142. https://doi.org/10.1109/TKDE.2004.47
40. Weijters, A., & Ribeiro, J. T. S. (2011). Flexible heuristics miner (FHM). In *2011 IEEE Symposium on Computational Intelligence and Data Mining (CIDM)*, (pp. 310–317). IEEE. https://doi.org/10.1109/CIDM.2011.5949453.
41. Aalst, W. M., De Medeiros, A. A., & Weijters, A. J. (2005). Genetic process mining. In *Applications and Theory of Petri Nets 2005: 26th International Conference, ICATPN 2005, Miami, USA, June 20-25, 2005. Proceedings* (Vol. 26, pp. 48–69). Springer. https://doi.org/10.1007/11494744_5
42. Medeiros, A. K. A., Weijters, A. J., & Aalst, W. M. (2007). Genetic process mining: An experimental evaluation. Data mining and knowledge discovery **14**, 245–304 (2007). https://doi.org/10.1007/s10618-006-0061-7
43. Tax, N., Verenich, I., La Rosa, M., Dumas, M. (2017). Predictive business process monitoring with LSTM neural networks. In *Advanced Information Systems Engineering: 29th International Conference, CAiSE 2017, Essen, Germany, June 12-16, 2017, Proceedings* (Vol. 29, pp. 477–492), Springer. https://doi.org/10.1007/978-3-319-59536-8_30
44. Rama-Maneiro, E., Vidal, J., & Lama, M. (2021). Deep learning for predictive business process monitoring: Review and benchmark. *IEEE Transactions on Services Computing*. https://doi.org/10.1109/TSC.2021.3139807
45. Bukhsh, Z. A., Saeed, A., & Dijkman, R. M. (2021). Processtransformer: Predictive business process monitoring with transformer network. arXiv:2104.00721
46. Mans, R., Schonenberg, H., Leonardi, G., Panzarasa, S., Cavallini, A., Quaglini, S., & Van Der Aalst, W. (2008). Process mining techniques: An application to stroke care. In *MIE* (Vol. 136, pp. 573–578). https://doi.org/10.3233/978-1-58603-864-9-573
47. Nguyen, A., Chatterjee, S., Weinzierl, S., Schwinn, L., Matzner, M., & Eskofier, B. (2021). Time matters: Time-aware LSTMs for predictive business process monitoring. In *Process Mining Workshops: ICPM 2020 International Workshops, Padua, Italy, October 5–8, 2020, Revised Selected Papers* (Vol. 2, pp. 112–123). Springer. https://doi.org/10.1007/978-3-030-72693-5_9
48. Kurniati, A. P., Hall, G., Hogg, D., & Johnson, O. (2018). Process mining in oncology using the mimic-iii dataset. In *Journal of Physics: Conference Series* (Vol. 971, p. 012008). IOP Publishing. https://doi.org/10.1088/1742-6596/971/1/012008
49. Lull, J. J., Cid-Menéndez, A., Ibanez-Sanchez, G., Sanchez, P. L., Bayo-Monton, J. L., Traver, V., & Fernandez-Llatas, C. (2021). Interactive process mining applied in a cardiology outpatient department. In *International Conference on Process Mining* (pp. 340–351). Springer International Publishing Cham. https://doi.org/10.1007/978-3-030-98581-3_25
50. Trčka, N., Pechenizkiy, M., & Aalst, W.: Process mining from educational data. In *Handbook of educational data mining* (pp. 123–142) (2010)
51. Romero, C., Cerezo, R., Bogarín, A., & Sánchez-Santillán, M. (2016). Educational process mining: A tutorial and case study using Moodle data sets. In *Data mining and learning analytics: Applications in educational research* (pp. 1–28). https://doi.org/10.1016/10.1002/9781118998205.ch1
52. de AR Goncalves, J. C., Santoro, F. M., & Baiao, F. A (2009). Business process mining from group stories. In *2009 13th International Conference on Computer Supported Cooperative Work in Design* (pp. 161–166). IEEE. https://doi.org/10.1109/CSCWD.2009.4968052

53. Ingvaldsen, J. E., Gulla, J. A., Su, X., & Rønneberg, H. (2005). A text mining approach to integrating business process models and governing documents. In *On the Move to Meaningful Internet Systems 2005: OTM 2005 Workshops: OTM Confederated Internationl Workshops and Posters, AWeSOMe, CAMS, GADA, MIOS+ INTEROP, ORM, PhDS, SeBGIS, SWWS, and WOSE 2005, Agia Napa, Cyprus, October 31-November 4, 2005. Proceedings* (pp. 473–484). Springer https://doi.org/10.1007/11575863_67
54. Prabhu, S. M., & Raja, M. (2023). New architecture to facilitate the expansion of e-government. In *Recent advances in data and algorithms for e-Government* (pp. 55–86). Springer. https://doi.org/10.1007/978-3-031-22408-9_3
55. Mans, R. S., Schonenberg, M., Song, M., Aalst, W. M., & Bakker, P. J. (2009). Application of process mining in healthcare–a case study in a Dutch hospital. In *Biomedical Engineering Systems and Technologies: International Joint Conference, BIOSTEC 2008 Funchal, Madeira, Portugal, January 28-31, 2008 Revised Selected Papers* (Vol. 1, pp. 425–438). Springer. https://doi.org/10.1007/978-3-540-92219-3_32
56. Prabhu, S. (2019). Development of network-based framework for managing social information. PhD thesis, Ph. D. Thesis, Visvesvaraya Technological University
57. Smith, E. (1996). On the border of causality: Contact and confusion. *Theoretical Computer Science, 153*(1–2), 245–270. https://doi.org/10.1016/0304-3975(95)00123-9
58. Jain, R., Durresi, A., & Babic, G. (1999). Throughput fairness index: An explanation. In *ATM forum contribution* (Vol. 99). https://doi.org/www.cse.wustl.edu/~jain/atmf/ftp/atm99-0045.pdf
59. DeLone, W. H., & McLean, E. R.: The DeLone and McLean model of information systems success: A ten-year update. *Journal of Management Information Systems, 19*(4), 9–30. https://doi.org/10.1080/07421222.2003.11045748
60. Mardis, M. A., Hoffman, E. S., & Marshall, T. E. (2008). A new framework for understanding educational digital library use: Re-examining digital divides in US schools. *International Journal on Digital Libraries, 9*, 19–27. https://doi.org/10.1007/s00799-008-0035-z
61. Alkraiji, A. I. (2020). Citizen satisfaction with mandatory e-government services: A conceptual framework and an empirical validation. *IEEE Access, 8*, 117253–117265. https://doi.org/10.1109/ACCESS.2020.3004541
62. Welch, E. W., Hinnant, C. C., & Moon, M. J. (2005). Linking citizen satisfaction with e-government and trust in government. *Journal of Public Administration Research and Theory, 15*(3), 371–391 (2005). https://doi.org/www.jstor.org/stable/3525668
63. Swaid, S. I., & Wigand, R. T. (2009). Measuring the quality of e-service: Scale development and initial validation. *Journal of Electronic Commerce Research, 10*(1), 13–28. https://doi.org/www.jecr.org/sites/default/files/10_1_p02.pdf
64. Reason, P., & Torbert, W. (2001). The action turn: Toward a transformational social science. *Concepts and Transformation, 6*(1), 1–37. https://doi.org/10.1075/cat.6.1.02rea
65. IKSFDP2022—22 Nov—Inauguration—8.27-@11.00. Retrieved 22, December 2023, from https://youtu.be/KnyYL389o1A@8.27-@11.00
66. Prabhu, S. M. How to run an institution well. LinkedIn.com. Retrieved 22 December, 2023, from https://www.linkedin.com/pulse/how-run-institution-well-dr-shreekanth-m-prabhu/?trackingId=MRmvfOnyT0yJFJBilHuTuQ%3D%3D
67. Rojas, E., Munoz-Gama, J., Sepúlveda, M., & Capurro, D. (2016). Process mining in healthcare: A literature review. *Journal of Biomedical Informatics, 61*, 224–236. https://doi.org/10.1016/j.jbi.2016.04.007
68. Deming, A. R. D. (1991) *The American who taught the Japanese about quality*. New York: Fireside Simon and Schuster. https://doi.org/books.google.co.in/books/about/Dr_Deming.html?id=834YItvV9CIC&redir_esc=y
69. Anderson, J. C., Rungtusanatham, M., & Schroeder, R. G. (1994). A theory of quality management underlying the deming management method. *The Academy of Management Review, 19*(3), 472. https://doi.org/10.2307/258936

Chapter 3
Thinking Inside the Sandbox: Beyond Public Services Digitalization with Co-Production

Polyana Batista da Silva and Fabrício Ramos Neves

Abstract This chapter explores how Regulatory Sandboxes can promote innovation and transform public services for citizens through co-production. They can be centered on citizens' interests and rights and aim to enhance participation, inclusion, and personal data protection in the public sector. We argue that Regulatory Sandboxes are adaptable and forward-thinking regulatory solutions and frameworks that can be useful beyond digital transformation to regulate a range of issues such as fake news, privacy invasion, discriminatory biases, and the use of Artificial Intelligence for governments. Regulatory Sandboxes allow for rapid experimentation, iteration, and learning in a controlled environment, free from the burden of existing regulations. Citizens could be involved in this process by providing input on the types of innovations they would like to see tested in the Sandbox and participating in the testing process. Using relevant literature and case examples, we showcase the advantages and disadvantages of co-production practices, highlighting some initiatives that improved public services. The chapter provides insights into the benefits and trials of using Regulatory Sandboxes as a form of co-production to support policymakers and public managers in designing and implementing more effective public services that meet citizens' needs.

Keywords Regulatory sandboxes · Co-production · Public services · Citizens' needs

P. B. da Silva (✉)
School of Economics, Business Administration and Accounting at Ribeirão Preto, University of São Paulo, Ribeirão Preto, Brazil
e-mail: polyanasilva@usp.br

F. R. Neves
Federal Institute of Education, Science and Technology Baiano, Guanambi, Brazil

School of Arts, Sciences and Humanities, University of São Paulo, São Paulo, Brazil

F. R. Neves
e-mail: fabricioneves@alumni.usp.br

C. Gaie and M. Mehta (eds.), *Transforming Public Services—Combining Data and Algorithms to Fulfil Citizen's Expectations*, Intelligent Systems Reference Library 252, https://doi.org/10.1007/978-3-031-55575-6_3

3.1 Introduction

The financial crisis of 2008, marked by the collapse of the subprime mortgage market and the bankruptcy of Lehman Brothers, exposed critical vulnerabilities and inefficiencies in the financial system. Issues like excessive leverage, lack of transparency, misaligned incentives, and systemic risk came to the forefront, challenging the prevailing positive attitude toward financial innovation. As the crisis unfolded, two key regulatory mandates gained prominence: consumer protection, particularly for retail clients, investors, and depositors, and financial stability in the macroprudential context. In response to the weaknesses in regulation highlighted by the crisis, a significant process of reregulation ensued [1, 2]. It is within this context of post-crisis regulatory change that Financial Technology (Fintech) emerged.

Years later, the COVID-19 pandemic in 2020 also disrupted the financial sector, prompting a partial shutdown of key segments, particularly the service sector [3]. Simultaneously, the pandemic accelerated the digital transformation and adoption of innovative technologies in finance, including fintech, blockchain, and artificial intelligence. While these innovations offered efficiency, inclusion, and competition benefits, they also introduced new challenges like cyberattacks, data breaches, fraud, and consumer protection issues. The pandemic underscored the imperative for regulatory flexibility, innovation, and increased collaboration among stakeholders, including regulators, innovators, consumers, and industry associations. In response to these evolving needs and challenges, the concept of a regulatory sandbox emerged.

To illustrate what a Regulatory Sandbox is, picture a studio for avant-garde financial artists. This studio is called a Regulatory Sandbox. Startups are like painters in this imaginative space who create their masterpieces by experimenting with their bold strokes of innovation, while regulators serve as discerning art critics, guiding and evaluating each brushstroke. Imagine a fintech startup, "CanvasFintech" (fictitious name), as a visionary painter creating a groundbreaking decentralized financial masterpiece. Within the Sandbox, CanvasFintech can layer its artistic strokes under the watchful eye of regulatory curators, ensuring that the final artwork adheres to the gallery's rules and does not splatter unintended risks. The Sandbox allows CanvasFintech to refine its masterpiece like an artist adjusting colors and textures. Once the artwork harmonizes innovation with regulatory harmony, the regulators (acting as gallery directors) grant CanvasFintech the privilege of unveiling its visionary creation to the broader financial art world, confident that it represents both artistic ingenuity and regulatory integrity.

Regulatory Sandboxes are environments where businesses—mostly those in the Financial Technology sector—can test innovative products, services, models, or delivery mechanisms in a controlled regulatory environment. Designed to strike a delicate balance between encouraging innovation and upholding regulatory standards, these sandboxes provide a structured environment for businesses to push the boundaries of their creativity [1]. This chapter explores the exciting potential of Regulatory Sandboxes in promoting innovation and transforming public services through co-production, an open innovation paradigm that actively involves citizens in the innovation cycle of cities or governments [4]. We focus on a specific breed of sandboxes—those influencing public services and innovation through co-production. In recent years, public administrations have moved beyond simply digitizing administrative forms and have started redesigning entire processes and services [5]. Embracing co-production approaches, they involve diverse stakeholders in transformative developments to elevate provision quality and generate public value [4, 6].

Regulatory Sandboxes have emerged as incubation models that facilitate the development of startups and foster innovation in financial markets [7] but ended up gaining prominence in several other areas. These sandboxes serve as regulatory frameworks that allow qualified candidates to experiment with their technology-enabled business solutions within a specified time and subject to certain conditions as stipulated by regulatory bodies [8]. Additionally, by using sandboxes, regulators can learn about new technologies and innovations and create suitable regulations to control risks [1]. Despite their significance, little attention has been given to studying the operational aspects of Sandboxes as a unique incubation and co-production model.

In a world full of startups, many obstacles can impede progress, such as high compliance expenses and limited regulatory knowledge. These challenges can lead to business failure and market disruption. To overcome these hurdles, governments have set up specialized Business Intelligence programs and accelerators that grant startups access to resources, substantially lowering validation costs and time [9]. For example, the US government's National Incubator Initiative of Clean Energy and the Brazilian Central Bank's regulatory sandboxes and innovation centers in the financial industry [10–12]. This rising focus on backing industry-specific incubator groups demonstrates a greater commitment to encouraging innovation within regulated sectors.

Various regulatory authorities worldwide, including the UK, US, Singapore, Australia, and Latin American countries, have recognized the importance of implementing effective regulatory solutions to foster innovation, enhance market competition, and maintain market stability. In fact, the UK's Financial Conduct Authority pioneered the creation of Regulatory Sandboxes in 2016, inspiring other governments to develop their own [13].

Regulatory Sandboxes stand out among other incubation models because they provide regulatory guidance and facilitate cross-jurisdictional testing [7]. FinTech startups utilize digital technologies such as blockchain, big data analytics, and artificial intelligence to create innovative services like digital payment solutions that disrupt traditional banking. While striving to enhance participation and inclusion in the public sector, Regulatory Sandboxes prioritize citizens' interests, rights, and data protection. Thus, citizens can actively contribute by providing input on the innovations they would like to see tested and participating in the testing process.

Despite the importance of creating and operating sandboxes for stimulating innovation and minimizing market disruptions, there is a lack of academic research that examines the operational aspects of sandboxes from a management point of view.

So, **how can co-production within Regulatory Sandboxes help to drive innovation and transform public services for citizens?** Existing studies predominantly focus on legal issues while neglecting other critical aspects, such as integrating co-production into regulatory sandboxes, which could lead to more effective and efficient public services [8, 14, 15].

Co-production allows citizens to actively contribute to constructing innovative products and services that serve their needs. Regulatory Sandboxes can utilize co-production to encourage citizen engagement in innovation development, ensuring that new products and services meet citizens' needs by involving them in the development process. Co-production can be achieved through stakeholder engagement, public feedback, co-creation, citizen participation, and transparency. It is a crucial element of citizen involvement in innovation, as it guarantees that the products and services created are relevant and beneficial to citizens.

We address this gap by bringing the co-production literature through propositions extending the understanding of this unique incubation model. We argue that Regulatory Sandboxes are adaptable and forward-thinking regulatory solutions and frameworks that can be useful beyond digital transformation to regulate a range of issues such as fake news, privacy invasion, discriminatory biases, and the use of Artificial Intelligence for governments. They allow rapid experimentation, iteration, and learning in a controlled environment with reduced regulatory burden. This chapter aims to explore the most effective practices of Regulatory Sandboxes and examine initiatives that utilize co-production to enhance public services by drawing on relevant literature and empirical examples and highlighting the benefits and drawbacks of these practices. We aim to suggest valuable insights into the benefits and tasks of utilizing Regulatory Sandboxes as a form of co-production, guiding policymakers and public managers in designing and implementing more effective public services aligned with citizens' needs. By examining the role of Regulatory Sandboxes in fostering improvements and transforming public services through co-production, this chapter contributes to the existing literature on incubation models and their management aspects. It sheds light on the operational dynamics of Regulatory Sandboxes, highlighting their potential to drive innovation, improve market competition, and enhance public services. Moreover, it emphasizes the meaning of citizen contribution and co-production in Regulatory Sandboxes, emphasizing the need for inclusive and participatory approaches to ensure the development of solutions that address citizens' needs effectively.

This chapter will first explore the current literature and real-world examples of Regulatory Sandbox models and discuss their relationship to co-production. Next, we will examine how these models can be expanded to drive innovation and improve public services on a larger scale. Our analysis results in propositions, and we will address some practical implications. Finally, we conclude with thoughts on future research opportunities.

3.2 Scaling Regulatory Sandboxes: Driving Innovation and Transforming Public Services

Regulatory Sandboxes, like versatile tools in the innovation toolkit, can be crucial in promoting innovation and transforming public services on a larger scale. These Sandboxes are typically located within countries and are limited in their range. Exploring effective strategies for scaling up Regulatory Sandboxes is essential to fully realize their potential. This will require careful consideration of the unique challenges and opportunities presented by different contexts. Ultimately, the successful expansion of regulatory sandboxes will depend on a shared vision and a commitment to working together to create a more innovative, responsive, and effective public sector.

First and foremost, Regulatory Sandboxes have traditionally been associated with the finance industry and fintech firms. However, their reach has expanded significantly, encompassing many sectors such as transport, energy, health, and Information and Communication Technology. It is as if these Sandboxes have spread their wings, soaring beyond their initial boundaries and venturing into new domains. For instance, Leckenby [16] demonstrated using a Sandbox in the UK's health sector. This expansion opens avenues for a wider range of innovative ideas and solutions, which can ultimately drive innovation and transform public services on a larger scale.

Typically, Regulatory Sandboxes are designed as small-scale programs, focusing on a limited number of firms that receive personalized guidance, support, and monitoring [17]. Nevertheless, what if we could bring more players into the Sandbox? Increasing the number of firms participating in Regulatory Sandboxes can be considered inviting more guests to a collaborative party. The larger the gathering, the more diverse the perspectives and expertise. This expanded participation injects dynamism into the Sandbox environment, leading to a more significant impact in promoting innovation and transforming public services on a larger scale.

Just like a well-constructed framework, developing a standardized approach ensures consistency and coherence. This involves establishing clear standards for the design and management of Regulatory Sandboxes. It is akin to establishing a common language that all participants understand. Moreover, developing a shared set of metrics for evaluating the impact of Sandboxes allows for effective comparisons and learning from different sandbox experiences. This standardized approach acts as a North Star, promoting innovation and transformation in public services on a larger scale.

Although Regulatory Sandboxes are valuable spaces for experimentation, their limited duration can be a barrier [8, 18]. However, imagine if we could extend this time frame. By providing firms more time in the Sandbox, we can offer them a nurturing environment to experiment, refine, and perfect their innovative products and services. It is similar to giving artists a larger canvas, allowing them the freedom to express their creativity fully. With an extended duration, firms could overcome challenges, fine-tune their offerings, significantly promote innovation, and transform public services on a larger scale.

From a regulator's standpoint, Sandboxes relies upon developing new, innovative, and adaptable approaches in addition to fintech. Technology-enabled solutions such as Regtech and Suptech are emerging, offering innovative solutions that could revolutionize the regulatory landscape [19, 20]. Regtech, short for Regulatory Technology, is a prime example of how innovative technologies can assist institutions in meeting compliance and regulatory reporting requirements. Its goal is to streamline and automate various processes associated with regulatory obligations, including data monitoring, reporting, and analysis. By leveraging Regtech, organizations can enhance their efficiency and accuracy in complying with regulations [19].

On the other hand, Supervisory Technology—Suptech, while closely related to Regtech, takes a different approach. It empowers supervisory agencies by equipping them with innovative technologies to sustain oversight and digitization of regulatory processes [19]. Suptech enables regulators to enhance their risk monitoring and compliance within financial institutions. Using data visualization, trend analysis, and forecasting, Suptech solutions improve the value and usability of the collected data, ultimately facilitating more effective regulatory decision-making.

In contrast, fintech encompasses a broader spectrum of technologies that disrupt traditional financial services. While fintech focuses on leveraging technology to enhance various aspects of financial services, including payments, lending, investments, and customer experience, Regtech specifically concentrates on regulatory compliance [20]. It provides specialized tools and solutions to help organizations navigate the complex web of regulatory requirements and ensure adherence.

To illustrate the practical application of Regtech, let us consider the case of Santander, a prominent financial institution. Santander has implemented the ABACUS/Transactions Regtech solution to meet the Money Market Statistical Reporting guidelines laid out by the European Central Bank. This cutting-edge solution has enhanced Santander's reporting procedures, enabling them to efficiently submit accurate and timely data to the European Central Bank and fulfill their regulatory obligations [19].

The application of Regulatory Sandboxes emerges as a transformative strategy, ushering in a new era for government services and citizens' experiences. We delve into two concrete examples at different implementation phases to elucidate these transformative benefits. Estonia, recognizing the imperative to attract foreign investment and talent, strategically harnessed a regulatory sandbox to refine its e-residency[1] program [21]. This approach enabled real-time user feedback, refined the program's functionality and user experience, and fostered collaboration among various government agencies involved in the e-Residency program. The result was a more efficient and coordinated approach, with the program providing global citizens access to Estonia's electronic services. Since its launch in 2014, it has proven immensely successful, boasting over 110,000 e-residents from 180 countries.

Expanding the scope of Regulatory Sandboxes into a diverse context, Rio de Janeiro initiated the 'Sandbox.Rio'[2] project in late 2022, marking it as the first

[1] https://www.e-resident.gov.ee/.

[2] https://www.sandboxrio.com.br/.

Brazilian capital to implement an environment dedicated to innovative solutions. This ambitious project, focused on optimizing government activities and promoting Smart City solutions, underscores the significance of a controlled testing environment. 'Sandbox.Rio' aims to enhance the municipal regulatory framework's receptiveness to new technologies and prioritize citizens' safety and urban coexistence. Furthermore, the Public Administration stands to benefit by refining public policies and crafting regulations based on real-time data collected within the experimental regulatory environment.

The project's inclusivity is evident in its diverse range of eligible initiatives, ranging from simplifying urban project licensing to promoting the digitalization of municipal public services, ensuring safe drone usage, installing structures for sustainable initiatives on municipal public roads, and introducing Smart City and Mobility as a Service (MaaS) solutions.

The concrete results remain uncertain while the 'Sandbox.Rio' project is still being implemented. Nonetheless, completing the selection process for innovative solutions holds promise for generating efficiency gains, reducing costs, and lowering entry barriers for innovative products, processes, and services within the city's business environment. The commitment to subjecting all projects to a technical, regulatory impact assessment underscores dedication to evaluating the relevance of regulating each product, striking a delicate balance between innovation and regulatory considerations. A central challenge for the project lies in establishing a dialogue with various public and private stakeholders involved in the initiative.

These tangible examples vividly underscore how regulatory sandboxes catalyze innovation, creating substantial value for government services and citizens. By providing a secure space for experimentation and collaboration, sandboxes facilitate the rapid development of solutions that enhance the efficiency, effectiveness, and accessibility of public services. The overarching goal is to usher in a better experience for all involved, fundamentally transforming traditional paradigms in government service delivery.

3.3 Bridging the Gap: Enhancing Project Lifecycles

Let's briefly draw parallels between the Regulatory Sandbox approach and the traditional project management lifecycle. The classic project lifecycle, a framework in project management, delineates stages from initiation to closure (Fig. 3.1), typically involving initiation, planning, execution, and closure phases [22, 23]. These phases provide structure but may lack the adaptability and real-time adjustments seen in Regulatory Sandboxes.

Organizational and legal regulations in the public sector can pose challenges to the procurement process. This may include limitations on budgeting schedules, supplier selection criteria, and information sharing. Such strict practices can particularly affect agile ICT projects, requiring more flexibility and technological capabilities than formal control [24, 25].

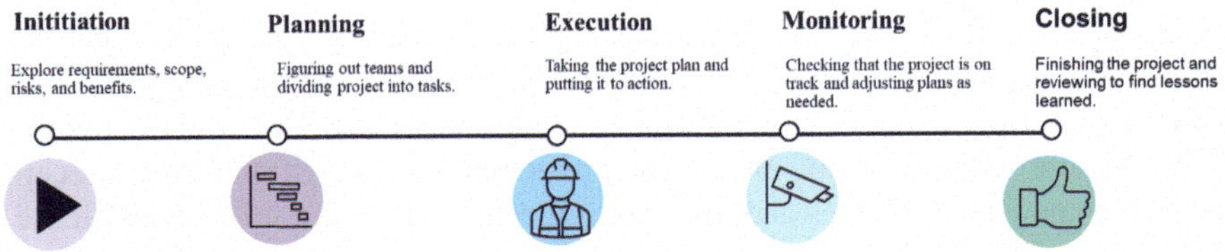

Fig. 3.1 Classical Project Management Lifecycle

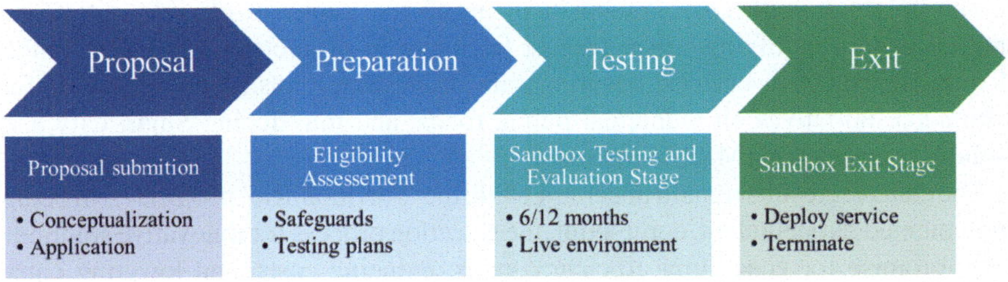

Fig. 3.2 Sandbox Lifecycle (adapted from Malta Financial Services Authority)

By examining sandboxes in the regulatory literature, it becomes evident that these tools include normative outlooks and technical approaches established over the last few decades [26]. The project lifecycle of a sandbox differs from that of a classical project (Fig. 3.2), as it depends on the context and introduces variability. For example, the UN DESA[3] Policy Brief outlines a sandbox cycle with stages like conceptualization, operation, testing, and evaluation or exit. Similarly, the Malta Financial Services Authority's FinTech Regulatory Sandbox[4] encompasses Conceptualization and Application, Preparation and Selection, Sandbox Testing, and Evaluation and Exit.

Taking them as a reference, in the preparation phase, participants collaborate closely to build safeguards and testing plans. The testing phase provides a live environment, with regulators offering customized guidance and monitoring compliance. Upon exiting, the participant may choose to deploy or terminate the service. This dynamic process contrasts the linear nature of the classic project lifecycle.

Sandboxes provide a controlled environment for testing new and innovative ideas without the full burden of existing regulations [18]. This allows for rapid experimentation, iteration, and learning, accelerating the development of impactful solutions. The sandbox project lifecycle generally emphasizes experimentation, testing, and evaluation more than the classic project lifecycle. It introduces safeguards, reporting requirements, and protocols for effective sandbox utilization. This iterative approach allows real-time adjustments, fostering reduced risk, increased efficiency, improved collaboration, and flexibility in managing project timelines and resources.

---

[3] https://www.un.org/development/desa/dpad/publication/un-desa-policy-brief-123-sandboxing-and-experimenting-digital-technologies-for-sustainable-development/.

[4] https://www.mfsa.mt/fintech/regulatory-sandbox/.

This dynamic collaboration fosters a culture of innovation that can transcend the Sandbox, influencing classical government services. As the private sector pioneers groundbreaking ideas within the Sandbox, the public sector gains exposure to innovative methodologies, emerging technologies, and new service models, setting the stage for transformative changes in the broader government landscape.

Recognizing the evident advantages of sandbox project lifecycles, it is crucial to acknowledge potential challenges, such as managing multiple sandboxes and coordinating with production environments, viewing these obstacles as opportunities for growth and learning. In deciding whether to adopt a sandbox project lifecycle, organizations must carefully consider these factors, balancing the benefits against potential disadvantages to create a project management environment that merges the classic lifecycle's structure with the sandbox approach's adaptability and innovation.

This integration aims to foster positive change and evolution in public services. The Sandbox's iterative nature promotes a dynamic and responsive approach, unlike the linear classic lifecycle where changes post-phase are often challenging to implement. The continuous testing and evaluation in the Sandbox allow for real-time adjustments, facilitating innovation and adaptation [7]. Ultimately, as illustrated here, the sandbox project lifecycle can enhance traditional project methodologies, providing a more agile and citizen-centric framework for project evolution.

3.4 From Participation to Co-Production: Fostering Citizen Involvement in Regulatory Sandbox Innovation

Regulation plays a vital role in governance, becoming increasingly important as governance processes grow. The emphasis on regulatory responsiveness acknowledges that governance is multifaceted and intricate [27]. Traditionally, regulation has been viewed as a top-down, command-and-control approach, which appears to differ from the values of co-production that emphasize engagement with external parties and promote equality and reciprocity [28].

Co-production, a collaborative approach where diverse entities work together to create products, services, or content, takes various forms, including joint ventures in the media industry and partnerships involving service users, carers, and citizens to enhance public services. In the public services sector, co-production involves equal collaboration to create services or make decisions that benefit all stakeholders, with active involvement from citizens and key stakeholders in developing and delivering public services and technology.

When examining regulation through the co-production lens, we uncover how regulatory influence seamlessly integrates into the co-production process. Collaborative arrangements involve partners with broad goals, but the co-production process shapes specific outcomes and methods, with regulatory influences typically blended or assimilated into the working processes [28].

In this context, the concept of Regulatory Sandboxes emerges as a promising opportunity to foster innovation and improve public services through collaboration. By involving citizens in shaping and implementing services, we can significantly enhance their quality and effectiveness [29]. It is essential to recognize varying levels of citizen involvement, ranging from passive participation, where the public is informed or manipulated without a significant role in decision-making, to co-creation, an active and collaborative approach in introducing, designing, or producing public services and policies [30]. This co-creative approach establishes a partnership where citizens have equal rights and responsibilities, often termed "enhanced participation" [31]. Within this collaborative framework, Regulatory Sandboxes offer a dynamic space for testing innovative ideas and solutions, ensuring that regulations evolve in tandem with the collaborative ethos of co-production.

A comprehensive definition of co-creation can be derived by combining insights from Torfing et al. [32] and Voorberg et al. [31]. According to this definition, co-creation involves co-initiation, co-design, and co-production, encompassing citizen initiatives and co-creation started by organizations. This definition provides a holistic understanding of the co-creation process, including the different stages of service delivery such as initiation, planning, and co-production. Daily processes and monitoring occur during co-production (see Table 3.1 below). Governments use many organizational mechanisms to support citizen initiatives [33]. Interestingly, co-initiation has often been overlooked in research on co-creation, yet it is a crucial element in public innovation processes co-creation [34].

Table 3.1 The Co-creation and Co-production process

The Co-creation/Co-production process				
Citizen as a co-initiator	Citizen as co-designer			Citizens actively participate in co-production
Citizen endeavor	Encourage citizen engagement	Understand everyone's roles and responsibilities	Ponding things together and producing a solution	Realize and put into action
At times, citizens might take the initiative while the government either provides support or becomes a hindrance	To promote co-creation, the government should inquire about effectively engaging its citizens	Provide information about the stakeholders involved and their contributions. How is civil society being engaged in this process?	What are the goals of working together?	Transform ideas, goals, and visions into tangible reality

Source Based on Voorberg et al. [31] and Torfing et al. [32]

It can be challenging to differentiate between co-creation and co-production since academics often use these terms interchangeably [31]. However, in the service management literature, co-creating value in collaboration with customers and promoting public value on a larger scale is emphasized [18, 35]. In contrast, in public administration literature, "co-production" is specifically used for public services [36–38]. Co-creation is closely linked to innovation and the development of groundbreaking ideas [32], and both approaches recognize the crucial position of citizens in public service delivery. To achieve successful co-creation and co-production, it is crucial to maintain diverse perspectives and foster synergy between the two [39].

In this vein, Perikangas and Tuurnas [40] suggest that design principles can be applied to enhance inclusion in what they call 'digital co-production' by addressing it at both the system level and during implementation. Considering the co-production model as a process for public governance could reveal its different impacts in public management contexts. Therefore, co-production holds the potential to empower citizens by enabling them to contribute their perspectives and preferences regarding the types of innovations they would like to see tested within the Sandbox. In addition, citizens can actively participate in the testing process, directly influencing the development of innovative solutions. Imagine a vibrant marketplace where everyone has a stake in the game, actively contributing their insights and knowledge. This cooperative approach might ensure that Regulatory Sandboxes effectively promote innovation, enhance regulatory standards, and protect consumer interests.

Acknowledging the significance of co-creation, the European Parliament highlights its significance in regulatory sandboxes to guarantee that innovative products and services are created with the citizens' needs in mind [41]. By embracing co-production, governments not only foster participation, inclusion, and personal data protection within the public sector but also establish an environment that addresses the needs and aspirations of citizens. Encouraging co-production, governments create an ecosystem where diverse voices converge, igniting innovation and driving the transformation of public services on a larger scale.

Viewing regulation as an integral aspect of co-production represents a crucial deviation from traditional interpretations of these concepts. While certain literature on co-production may touch on the topics of "guides" and "self-management," regulation is seldom mentioned [28]. The primary goal of Regulatory Sandboxes is to gain valuable insights into the opportunities and threats related to specific innovations and to develop the appropriate regulatory frameworks to support their safe and beneficial integration into society. This process ensures that any new products, services, or solutions arising from these innovations meet the necessary standards of consumer safety.

Within Regulatory Sandboxes, businesses are provided with a secure and responsible space to test their novel ideas, products, services, technologies, business models, and even policies. By exempting these businesses from burdensome regulations, governments encourage them to experiment and explore innovative possibilities freely. This freedom to test and refine ideas ultimately paves the way for groundbreaking advancements and transformative solutions [42]. Furthermore, regulatory sandboxes facilitate collaboration between businesses and regulators, fostering co-production as a core principle. By working together, these stakeholders can jointly develop innovative products and services that align with the needs and aspirations of citizens. This collaborative effort ensures that the resulting innovations are user-centric, delivering tangible benefits to the public [38].

By adapting or relaxing regulations within the Sandbox environment, governments create a conducive atmosphere for innovation to thrive. This flexibility in regulatory approaches allows for a more agile and adaptive system, fostering continuous progress and breakthroughs. Lastly, Regulatory Sandboxes has the potential to enable regulators to enhance their understanding of the opportunities and risks associated with emerging innovations. By gaining firsthand experience and insights, regulators can develop the necessary expertise to create appropriate regulatory environments that safeguard consumer interests. This proactive approach ensures that innovative products and services reach the market safely and responsibly.

3.5 Transforming Public Services: The Role of Regulatory Sandboxes and Co-Production in Fostering Innovation for Citizens

Limited examples show how Regulatory Sandboxes can achieve these goals when it comes to promoting innovation and transforming public services through a co-production approach. To illustrate the potential of such arrangements, we have carefully selected a few cases to delve into, demonstrating how a dedicated and secure architecture can be defined and implemented through concrete use cases.

One notable example is the CGAP Financial Innovation Sandbox [43], established by a financial sector regulator in collaboration with the Consultative Group to Assist the Poor, a think tank consortium in the World Bank. This Sandbox provides a regulated environment for organizations to live-test small-scale technologies. It aims to encourage innovation, improve regulatory standards, and protect and serve impoverished individuals. The Sandbox operates through co-production between stakeholders, ensuring that the innovative products and services meet the needs of consumers. By extending digital financial services, the Sandbox contributes to financial inclusion.

Another initiative worth mentioning is the work of Engels et al. [44], who conducted a thorough analysis of two case studies in Germany. These cases applied the concepts of test beds and living labs. Testbeds differ from traditional experiments in society by being experiments on society. Testbeds are constructed with a regulatory focus to reduce local barriers, often in "sandbox" settings, and test innovative technologies with uncertain outcomes. Regulatory sandboxes provide transparent environments for testing innovative products and services using generic and specialized incubation models.

A study conducted by Gasco [45] in Spain analyzed two case studies—a public fab labs' network in Barcelona and the Citilab in Cornellà City. The findings showed that living labs are useful in co-developing technology and regulatory frameworks. Living labs encourage open innovation, fostering collaboration in research, development, and real-world experimentation. Citizens play a crucial role in this process. To attain their objectives, living labs rely on specific methods and techniques implemented through innovative projects and community-building initiatives.

In the Netherlands, the Dutch Green Deal promotes and supports innovative 'green' initiatives [46]. Aiming to facilitate sustainable initiatives, a thematic approach encompassed nine areas: climate, raw materials, bio-based economy, construction, mobility, biodiversity, food, water, and energy. Barriers were removed, information and advice provided, partnerships enabled, and new markets created without focusing on specific technology or sectors. The goal was to encourage comparable projects through spin-offs and spillover outcomes attained through mutual agreements or 'covenants' signed by a coalition of participants under Dutch private law.

In their study on Regulatory Sandboxes for autonomous vehicles in Singapore, Tan and Taeihagh [47] discovered that successful experimentation is facilitated by government stewardship and collaboration with stakeholders to address risks. Additionally, enhancing policy capacity and creating a supportive business environment were found to be crucial. However, the authors suggested that participatory approaches would enable governments to understand citizens' perspectives, preferences, and long-term visions for transportation and other large-scale plans. Such an approach would allow governments to tap into citizens' expertise and collaborate with them to develop solutions for the risks associated with the deployment of autonomous vehicles. Regulatory Sandboxes could play a role by issuing guidance for public feedback, ensuring that the entire market can benefit from citizens' input in shaping innovative products and services.

Individuals in South Korea can now use a Decentralized Identity system that operates through blockchain technology, thanks to successful development by the government. This system allows digitally certified identities to be managed within a Regulatory Sandbox. As per Rim's [48] research, the co-production of diverse actors and their pursuit of legitimacy greatly contribute to developing new information systems and creating public value. This innovative approach revolutionizes how organizations communicate with citizens in the digital economy.

Lastly, Undheim and colleagues [49] presented the case of the Norwegian regulatory Sandbox for responsible artificial intelligence, arguing that regulatory sandboxes can potentially support the development of a more ethical AI. This Sandbox was established in 2020 by the Data Protection Authorities in response to a government white paper on AI governance. It aims to promote developing and implementing ethical and responsible AI from a privacy perspective. Applications for the Sandbox are assessed and selected by an internal Data Protection Authority committee in collaboration with an external reference group to ensure social relevance and benefits. However, the authors did not show results from applying the Sandbox. Additionally, the authors call for further empirical research to understand the implications of uncertainty in creating ethical AI.

The same movement could be seen in Brazil. In late October 2023, the Brazilian Data Protection Authority[5] took a significant stride by initiating a call for contributions to its regulatory sandbox pilot program. This pioneering program introduces a controlled environment specifically tailored for testing technologies related to Artificial Intelligence (AI), with active participation from various stakeholders. The primary objective is to stimulate innovation within this controlled space, fostering the implementation of best practices to ensure compliance with personal data protection rules and principles. Notably, the Authority's regulatory sandbox pilot is a collaborative effort, shaped in conjunction with the Development Bank of Latin America and the Caribbean, leveraging shared methodologies, knowledge, and experiences to develop regulatory sandboxes in the region.

Through this consultation, the regulatory Sandbox aims to collect pertinent information and data, providing a comprehensive understanding of the perspectives of diverse stakeholders whom future regulations in the dynamic fields of Artificial Intelligence and data protection may impact. This initiative underscores the commitment to creating an inclusive and informed regulatory framework that considers the nuanced perspectives of various actors within the evolving landscape of AI and data protection. This is related to the work of Gonzalez Torrez and Sawhney [50], who proposed that AI regulatory sandboxes, in conjunction with integrated Machine Learning Operations processes and practices, can offer crucial mechanisms for regulatory experimentation and technological innovation with multiple stakeholders across the lifecycle of AI-enabled services in the public sector.

The examples showcase how Regulatory Sandboxes can promote innovation and transform public services by involving businesses, regulators, and citizens in the development process. These instances exemplify how sandboxes can foster collaborative solution creation, elevate regulatory standards, and ultimately improve the overall quality of public services for the greater good.

3.6 Practical Implications

In recent years, there has been a significant increase in the digitization of public services and the adoption of e-Government initiatives. This has led to a wave of innovation in public organizations that was never seen before. It is not just about the usual benefits of increased efficiency but unlocking the potential of Regulatory Sandboxes as dynamic catalysts that can transform the delivery of public services.

---

[5] https://www.gov.br/anpd/pt-br/assuntos/noticias/anpds-call-for-contributions-to-the-regulatory-sandbox-for-artificial-intelligence-and-data-protection-in-brazil-is-now-open.

The continuous decrease in computing and storage costs has significantly increased the effectiveness of Artificial Intelligence and Machine Learning tools. This has marked the beginning of a digital era where cloud computing is the foundation of data-driven innovation. Data has become incredibly important in economic and social sectors, blurring the boundaries between public and private domains. Due to the complexity of the Datasphere, a systems approach to data governance is necessary [19]. This is where Regulatory Sandboxes come into play. These experimental spaces provide an innovative way to iteratively address challenges, offering the flexibility required for complex systems.

We believe that Regulatory Sandboxes have a practical impact beyond conventional e-Government procedures. These sandboxes provide a platform for exploring various data types and promote responsible sharing, leading to improved service value and personalized experiences that exceed expectations. As demonstrated in the examples within this chapter, Regulatory Sandboxes can ensure cost efficiency, streamline processes, and intelligently allocate resources, thus driving savings and ensuring public funds are used effectively. Additionally, seamless service delivery reduces time by eliminating unnecessary bureaucratic hurdles, freeing up citizens' time and resources. A secure and reliable infrastructure guarantees data privacy and fosters trust in government services, resulting in elevated quality of IT service. Integrating these new services with conventional e-Government procedures can create synergy and amplify their collective impact on the citizen experience.

These sandboxes allow for exploring various data types and promote responsible sharing, resulting in improved service value and personalized experiences that exceed expectations. As demonstrated in the chapter's examples, Regulatory Sandboxes can ensure cost efficiency, streamline processes, and intelligently allocate resources, thus driving savings and ensuring public funds are used effectively. Additionally, seamless service delivery reduces time by eliminating unnecessary bureaucratic hurdles, freeing up citizens' time and resources. A secure and reliable infrastructure guarantees data privacy and fosters trust in government services, resulting in elevated IT service quality. The interconnection of these new services with conventional e-Government procedures can create synergy and amplify their collective impact on the citizen experience.

But we can go further. Imagine integrating blockchain technology within Regulatory Sandboxes to revolutionize data governance, ensuring security, privacy, and traceability. Collaborative Sandboxes operating across borders facilitate international cooperation, fostering a global community focused on responsible and innovative governance. Introducing incentives for sustainable innovations within Regulatory Sandboxes, could lead to fast-tracking approvals for eco-friendly solutions. And why not bring Virtual Reality platforms in? Imagine immersive public consultations where citizens can explore and provide feedback on proposed innovations, ensuring a more inclusive and participatory decision-making process. Finally, algorithmic impact assessments within Regulatory Sandboxes could be implemented to evaluate AI-powered innovations' ethical and societal implications. This might ensure that emerging technologies are efficient and adhere to ethical standards, preventing biases and discriminatory outcomes. These proposals push the boundaries, transforming Sandboxes into dynamic ecosystems for transformative solutions.

Governments should recognize the transformative potential of Regulatory Sandboxes and adopt a strategic approach to shaping them. The objectives of these Sandboxes must be clear, with a focus on promoting innovation, uplifting regulatory standards, and safeguarding citizen interests. These flexible frameworks are not limited to digital transformation but also regulate fake news, privacy invasion, discriminatory biases, and AI. It is crucial to adopt an inclusive approach to ensure that Regulatory Sandboxes evolve with a deep understanding of diverse stakeholder needs.

Guidance and support from governments are crucial in promoting innovation. Governments should provide information on legal possibilities, future competitions, and practical examples to foster a global exchange of innovative ideas. Rigorous evaluation should be conducted to identify areas for improvement and assess intended objectives. Transparency should be the linchpin, cultivating an environment where Regulatory Sandboxes can evolve into dynamic ecosystems.

When citizens are actively involved, their feedback is gathered on preferred innovations, and their participation is enabled in testing, Sandboxes may become powerful instruments for promoting innovation, elevating standards, and protecting consumers. This is not just a theoretical exercise. Embracing this vision can transform Regulatory Sandboxes into vibrant ecosystems, propelling us toward a future where public service delivery is efficient and transformative—a future that empowers citizens and shapes the trajectory of societal progress.

3.7 Conclusion

The concept of co-production and the innovative approach to regulation offer a promising solution to the challenges posed by over-regulation. There is still an opportunity to push the boundaries of traditional public services and ensure their adoption by citizens, especially in this transformative moment when the relationship between law, state, and society is undergoing significant change.

Thinking inside the Sandboxes requires seeing them as dynamic environments where individuals can engage in experimentation, dialogue, learning, and collaborative efforts, bringing a place for stakeholders from various sectors to come together, imagine, and shape technological advancements that benefit everyone involved. These stakeholders include private companies, data regulators, civil society organizations, academics, data subjects, and regulators from different jurisdictions.

We believe that Regulatory Sandboxes offer adaptable and forward-thinking regulatory solutions and frameworks that extend beyond digital transformation. They can be valuable tools for addressing a wide range of issues, including fake news, privacy invasion, discriminatory biases, and the use of AI by governments. They enable rapid experimentation and learning in a controlled environment without the burden of existing regulations. This allows for developing and testing innovative solutions that address user needs more efficiently and effectively. By leveraging Regulatory

Sandboxes as a form of co-production, policymakers and public managers can design and implement more effective public services that align with the needs of citizens.

The insights we presented here might shed light on the benefits and challenges associated with the use of Regulatory Sandboxes by governments. To successfully implement Regulatory Sandboxes, it is essential to establish clear objectives, engage all relevant actors, provide guidance and support, evaluate impact, and ensure transparency throughout the process. To fully maximize the potential of Sandboxes, governments should actively involve citizens, businesses, and government agencies. Together, these stakeholders can work towards creating a regulatory environment that embraces the complexities of the digital era and fosters innovation.

Like a sandbox where children explore, build, and experiment, Regulatory Sandboxes can supply a safe and controlled space for testing and refining innovative ideas. It brings together different stakeholders and fosters collaboration, enabling the development of effective solutions and the advancement of regulatory frameworks. Regulatory Sandboxes may act as catalysts for digital transformation, unleashing emerging technologies' potential and promoting improved public service delivery. This digital co-production has the potential to solve the obstacles associated with in-person co-production by transcending limitations of time and distance, aiding in citizen mobilization, and reducing expenses while promoting information exchange. This approach might provide a potential answer to the absence of traditional avenues for citizen engagement employed by governments and an opportunity to tackle the difficulties of conventional citizen commitment. Moreover, technological advancements have made co-production more feasible, inclusive, and transparent.

As we look to the future, several intriguing research avenues exist. A potential area of study is the long-term impact of Regulatory Sandboxes on innovation ecosystems and the broader regulatory landscape. Additionally, investigating the role of ethics and privacy considerations within Sandboxes can help ensure responsible innovation. Moreover, examining the transferability and scalability of successful sandbox models across different contexts and jurisdictions would provide valuable insights.

By further exploring these research avenues, we can enhance our comprehension of the capabilities of Regulatory Sandboxes and how they can shape the regulatory frameworks of tomorrow.

References

1. Zetzsche, D. A. et al. (2017). Regulating a revolution: From regulatory sandboxes to smart regulation. *Fordham J Corp & Fin L., 23*, 31–103. https://ir.lawnet.fordham.edu/jcfl/vol23/iss1/2/.
2. Allen, H. J. (2019). Regulatory sandboxes. *Geo Wash L Rev., 87*, 579
3. Putnam, B. H. (2021). From phase transitions to modern monetary theory: A framework for analyzing the pandemic of 2020. *Review of Financial Economics, 39*(1), 3–19. https://doi.org/10.1002/rfe.1122.

4. Scupola, A., & Mergel, I. (2022). Co-production in digital transformation of public administration and public value creation: The case of Denmark. *Government Information Quarterly, 39*(1). https://doi.org/10.1016/j.giq.2021.101650.
5. Mergel, I., Edelmann, N., & Haug, N. (2019). Defining digital transformation: Results from expert interviews. *Government Information Quarterly, 36*(4). https://doi.org/10.1016/j.giq.2019.06.002.
6. Tonurist, P. (2018). New approaches in policy design and experimentation. *OECD: OECD Science, Technology and innovation outlook 2018.* https://doi.org/10.1787/sti_in_outlook-2018-16-en.
7. Alaassar, A., Mention, A. L., & Aas, T. H. (2021). Exploring a new incubation model for FinTechs: Regulatory sandboxes. *Technovation, 103.* https://doi.org/10.1016/j.technovation.2021.102237.
8. Makarov, V. O., & Davydova, M. L. (2021). On the Concept of Regulatory Sandboxes. In E. G. Popkova, B. S. Sergi (Eds.), *Smart technologies for society, state and economy. ISC 2020. Lecture notes in networks and systems* (Vol. 155). Springer. https://doi.org/10.1007/978-3-030-59126-7_112.
9. Fenwick, M., Vermeulen, E. P. M. (2020). Fintech, overcoming friction and new models of financial regulation. In M. Fenwick, S. Van Uytsel & B. Ying (Eds.), *Regulating FinTech in Asia. Perspectives in law, Business and innovation.* Springer. https://doi.org/10.1007/978-981-15-5819-1_11.
10. Banco Central do Brasil (BCB). (2023). *Regulatory Sandbox.* https://www.bcb.gov.br/en/financialstability/regulatorysandbox. Accessed on 3 June 2023.
11. Neves, F. R., & Silva, P. B. (2023). From paper to digital: e-Government's evolution and pitfalls in Brazil. In C. Gaie, & M. Mehta (Eds.), *Recent advances in data and algorithms for e-Government. artificial intelligence-enhanced software and systems engineering* (Vol. 5). Springer. https://doi.org/10.1007/978-3-031-22408-9_9.
12. Herrera, D., Vadillo, S. (2018). Regulatory sandboxes in Latin America and the Caribbean for the fintech ecosystem and the financial system. *Inter-American Development Bank, IDB-DP, 573,* 1–41. https://cdn.crowdfundinsider.com/wp-content/uploads/2018/04/Regulatory-Sandboxes-in-Latin-America-and-the-Caribbean-for-the-FinTech-Ecosystem-and-the-Financial-System.pdf.
13. Authority, F. C. (2015). Regulatory sandbox. *Financial Conduct Authority, 11,* 30. https://www.ifashops.com/wp-content/uploads/2015/11/regulatory-sandbox.pdf. Accessed on 3 June 2023
14. Arner, D. W., Barberis, J., & Buckley, R. P. (2017). *FinTech and RegTech in a Nutshell, and the Future in a Sandbox.* CFA Institute Research Foundation. https://www.cfainstitute.org/research/foundation/2017/fintech-and-regtech-in-a-nutshell-and-the-future-in-a-sandbox.
15. Bromberg, L., Godwin, A., Ramsay, I. (2017). Fintech sandboxes: Achieving a balance between regulation and innovation. *Journal of Banking and Finance Law and Practice, 28*(4), 314–336. https://ssrn.com/abstract=3090844.
16. Leckenby, E., Dawoud, D., Bouvy, J., & Jónsson, P. (2021). The sandbox approach and its potential for use in health technology assessment: A literature review. *Applied Health Economics and Health Policy, 19,* 857–869. https://doi.org/10.1007/s40258-021-00665-1.
17. Alaassar, A., Mention, A. L., & Aas, T. H. (2023). Facilitating innovation in FinTech: A review and research agenda. *Review of Managerial Science, 17*(1), 33–66. https://doi.org/10.1007/s11846-022-00531-x.
18. Yefremov, A. (2019). Regulatory sandboxes and experimental legislation as the main instruments of regulation in the digital transformation. In D. Alexandrov, A. Boukhanovsky, A. Chugunov, Y. Kabanov, O. Koltsova, I. Musabirov (Eds.), *Digital transformation and global society. DTGS 2019. Communications in computer and information science* (Vol. 1038). Springer. https://doi.org/10.1007/978-3-030-37858-5_7.
19. Papantoniou, A. A. (2022). Regtech: Steering the regulatory spaceship in the right direction? *Journal of Banking and Financial Technology, 6*(1), 1–16. https://doi.org/10.1007/s42786-022-00038-9.

20. Grassi, L., & Lanfranchi, D. (2022). RegTech in public and private sectors: The nexus between data, technology and regulation. *Journal of Industrial and Business Economics, 49*(3), 441–479. https://doi.org/10.1007/s40812-022-00226-0.
21. Kerikmäe, T., & Pärn-Lee, E. (2021). Legal dilemmas of Estonian artificial intelligence strategy: In between of e-society and global race. *AI & SOCIETY, 36*, 561–572. https://doi.org/10.1007/s00146-020-01009-8.
22. Picciotto, R. (2020). Towards a 'New Project Management' movement? An international development perspective. *International Journal of Project Management, 38*(8), 474–485. https://doi.org/10.1016/j.ijproman.2019.08.002.
23. Morris, P. W. G. (2013). Reconstructing project management reprised: A knowledge perspective. *Project Management Journal, 44*(5), 6–23. https://doi.org/10.1002/pmj.21369.
24. De Schepper, S., Dooms, M., & Haezendonck, E. (2014). Stakeholder dynamics and responsibilities in Public-Private Partnerships: A mixed experience. *International Journal of Project Management, 32*(7), 1210–1222. https://doi.org/10.1016/j.ijproman.2014.01.006.
25. Wirick, D. (2011). *Public-sector project management: meeting the challenges and achieving results*. John Wiley & Sons.
26. Johnson, W. G. (2023). Caught in quicksand? Compliance and legitimacy challenges in using regulatory sandboxes to manage emerging technologies. *Regulation & Governance, 17*(3), 709–725. https://doi.org/10.1111/rego.12487.
27. Bourgon, J. (2007). Responsive, responsible and respected government: Towards a New Public Administration theory. *International Review of Administrative Sciences, 73*(1), 7–26. https://doi.org/10.1177/0020852307075686.
28. Innes, M., Davies, B., & McDermont, M. (2019). How co-production regulates. *Social & Legal Studies, 28*(3), 370–391. https://doi.org/10.1177/0964663918777803.
29. Attrey, A., Lesher, M., Lomax, C. (2020). *The role of sandboxes in promoting flexibility and innovation in the digital age*. https://doi.org/10.1787/cdf5ed45-en.
30. Lipp, B., Baudrin, M., Cuevas-Garcia, C., Pepponi, F., Rozwadowska, M., Tsui, S. (2023). Co-creating end-user roles. Understanding the new variety of user involvement in public sector innovation. *Science and Public Policy, 50*(1), 146–159. https://doi.org/10.1093/scipol/scac077.
31. Voorberg, W. H., Bekkers, V. J., & Tummers, L. G. (2015). A systematic review of co-creation and co-production: Embarking on the social innovation journey. *Public management review, 17*(9). https://doi.org/10.1080/14719037.2014.930505.
32. Torfing, J., Sørensen, E., & Røiseland, A. (2019). Transforming the public sector into an arena for co-creation: Barriers, drivers, benefits, and ways forward. *Administration & Society, 51*(5), 795–825. https://doi.org/10.1177/0095399716680057.
33. Bakker, J., Denters, B., Oude Vrielink, M., & Klok, P. J. (2012). Citizens' initiatives: How local governments fill their facilitative role. *Local Government Studies, 38*(4), 395–414. https://doi.org/10.1080/03003930.2012.698240.
34. Sørensen, E., & Torfing, J. (2018). Co-initiation of collaborative innovation in urban spaces. *Urban Affairs Review, 54*(2). https://doi.org/10.1177/1078087416651936.
35. Osborne, S. P., & Strokosch, K. (2013). It takes two to tango? understanding the co-production of public services by integrating the services management and public administration perspectives. *British Journal of Management, 24*, S31–S47. https://doi.org/10.1111/1467-8551.12010.
36. Nabatchi, T., Sancino, A., & Sicilia, M. (2017). Varieties of participation in public services: The who, when, and what of co-production. *Public Administration Review, 77*(5), 766–776. https://doi.org/10.1111/puar.12765.
37. Bovaird, T. (2007). Beyond engagement and participation: user and community co-production of public services. *Public Administration Review, 67*(5), 846–860. https://doi.org/10.1111/j.1540-6210.2007.00773.x.
38. Boxelaar, L., Paine, M., & Beilin, R. (2006). Community engagement and public administration: Of silos, overlays and technologies of government. *Australian Journal of Public Administration, 65*(1), 113–126. https://doi.org/10.1111/j.1467-8500.2006.00476.x.
39. Osborne, S. P., Radnor, Z., & Strokosch, K. (2016). Co-production and the co-creation of value in public services: A suitable case for treatment? *Public Management Review, 18*(5), 639–653. https://doi.org/10.1080/14719037.2015.1111927.

40. Perikangas, S., & Tuurnas, S. (2023). Design for inclusive digital co-production. *Public Management Review*, 1–21. https://doi.org/10.1080/14719037.2023.2224819.
41. Parenti, R. (2020). Regulatory sandboxes and innovation hubs for FinTech. *Study for the Committee on Economic and Monetary Affairs, Policy Department for Economic, Scientific and Quality of Life Policies, European Parliament, Luxembourg, 65*. https://www.europarl.europa.eu/RegData/etudes/STUD/2020/652752/IPOL_STU(2020)652752_EN.pdf.
42. Alaassar, A., Mention, A. L., & Aas, T. H. (2020). Exploring how social interactions influence regulators and innovators: The case of regulatory sandboxes. *Technological Forecasting and Social Change, 160*. https://doi.org/10.1016/j.techfore.2020.120257.
43. Jenik, I., & Lauer, K. *Regulatory sandboxes and financial inclusion*. CGAP. https://www.cgap.org/sites/default/files/researches/documents/Working-Paper-Regulatory-Sandboxes-Oct-2017.pdf. Accessed 30 June 2023
44. Engels, F., Wentland, A., & Pfotenhauer, S. M. (2019). Testing future societies? Developing a framework for test beds and living labs as instruments of innovation governance. *Research Policy, 48*(9). https://doi.org/10.1016/j.respol.2019.103826.
45. Gascó, M. (2017). Living labs: Implementing open innovation in the public sector. *Government Information Quarterly, 34*(1), 90–98. https://doi.org/10.10176/j.giq.2016.09.003.
46. Rosemberg, C., Potau, X., Dijkstal, F., Vinnik, A., Tiriduzzi, C., Daved, A., & Blind, K. (2020). *Regulatory sandboxes and innovation testbeds. a look at international experience and lessons for Latin America and the Caribbean*. https://doi.org/10.18235/0002526.
47. Tan, S. Y., & Taeihagh, A. (2021). Adaptive governance of autonomous vehicles: Accelerating the adoption of disruptive technologies in Singapore. *Government Information Quarterly, 38*(2). https://doi.org/10.1016/j.giq.2020.101546.
48. Rim, H. (2023). Decentralized identity (DID): New technology adoption and diffusion in South Korea. *Transforming Government: People, Process and Policy, 17*(2), 251–270. https://doi.org/10.1108/TG-11-2021-0189.
49. Undheim, K., Erikson, T., & Timmermans, B. (2023). True uncertainty and ethical AI: Regulatory sandboxes as a policy tool for moral imagination. *AI and Ethics, 3*(3), 997–1002. https://doi.org/10.1007/s43681-022-00240-x.
50. Gonzalez Torres, A. P., & Sawhney, N. (2023). Role of regulatory sandboxes and mlops for ai-enabled public sector services. *Rev Socionetwork Strat, 17*, 297–318. https://doi.org/10.1007/s12626-023-00146-y.

Chapter 4
Assessing Sustainable Development of e-Government Services Towards Citizen Satisfaction Through Continuous Improvement Approach

D. Vimala, S. Vasantha, and A. Shanmathi

Abstract Online services offered by the government to individuals, companies, and other organisations are known as e-Government services. Digital technologies are used to deliver public services in a way that is sustainable in terms of the economy, society, and environment. Making sure that these services are easily available, reasonably priced, and effective is crucial for their sustainability. The "continuous improvement approach" refers to the process of regularly assessing and enhancing the quality of digital government services. Through a continuous improvement approach, this work adds to the idea of a new electronic transformation that would assist in developing nations to provide sustainable e-Government services. To achieve this, the researchers considered five factors: a rigorously controlled process, efficient services, agile accessibility, cutting-edge technology, trust, and awareness. The researchers discovered that the proposed approach was advantageous to both the government and the users, increasing the sustainability of those services. Through a continuous improvement strategy, the researchers built a model to analyse the link between sustainable e-Government services and citizen satisfaction. The proposed model was evaluated through structured questionnaire surveys. 150 respondents participated in the surveys that are using e-Government services. How the continuous improvement approach has improved citizen satisfaction in various nations was validated through case study method and secondary data.

Keywords Citizen satisfaction · Continuous improvement approach · e-Government services

---

D. Vimala
School of Management Studies, Technology and Advanced Studies(VISTAS), Vels Institute of Science, Chennai, India

S. Vasantha (✉) · A. Shanmathi
Hinduja Housing Finance, Chennai, India
e-mail: vasantha.sms@velsuniv.ac.in

C. Gaie and M. Mehta (eds.), *Transforming Public Services—Combining Data and Algorithms to Fulfil Citizen's Expectations*, Intelligent Systems Reference Library 252, https://doi.org/10.1007/978-3-031-55575-6_4

4.1 Introduction

The accomplishment of good governance objectives can be significantly aided by new information and communication technology. This "e-Governance" can improve governance's effectiveness and efficiency while also providing additional advantages. But the majority of e-Governance projects experience two problems. Creating the six prerequisites for e-Governance is the first strategic concern for e-readiness. Consequently, a perspective that is change-oriented is presented.

Organisations of all types are now utilising the Continuous Improvement approach as their new management paradigm. The Continuous Improvement concept can be successfully utilised to realise significant gains in the organization's manufacturing performance in a highly competitive environment. It is abundantly clear from the analysis of the literature that CI practises can aid industrial groups in achieving their goals and gaining a competitive advantage.

The previous paradigm involved data processing to automate government operations through information technology (IT) [1]. The new model is the one in which data processing and data communication technology, assist and modify the exterior operations of government. All ICTs should be considered as part of e-Government,

- Business and citizen connections that are altering service delivery.
- Networking among and within NGOs to enable learning and coordinated action.
- Interconnections between and among communities

The focus expands to include e-citizens, e-services, and e-society in addition to merely some aspects of e-administration.

There are substantial distinctions between the terms "government" and "governance". The term "governance" refers to a larger range of regulatory acts that are often not regulated by the formal government, whereas "government" refers to the institution. Even while corporate law may include the government in this, some issues continue to fall outside of its purview. Consider corporate governance, the word used to describe the internal practises created by the private sector to promote stakeholder responsibility [2]. Governance is defined as the formal and informal systems and procedures that guide and regulate a group's overall behaviour. Government is the component of society that enacts legislation and wields power. Governments are required to perform their obligations independently. In the creation of governance, private enterprises, business organisations, nongovernmental organisations (NGOs), and associations of NGOs all participate, frequently collaborating with governmental institutions but occasionally acting independently of formal authority [3].

4.2 Objectives

- To analyse the various dimensions of sustainability in e-Government services.
- To analyse the influence of sustainability of e-Government services adoption on citizen satisfaction.
- To review literature on the role of continuous improvement approach towards citizen satisfaction.
- To discuss the role of Continuous Improvement Approach towards citizen satisfaction through case study method across various countries.

4.3 Research Methodology

The researchers developed a model to analyse the impact of sustainability in e-Government Services adoption over Citizen Satisfaction. A structured questionnaire was used to collect primary data to validate the conceptual model. Case studies and other secondary sources of data were examined in order to establish the relationship between Continuous Improvement Approach and Citizen Satisfaction. A single item, 5-point Likert type scale rated overall citizen satisfaction with Strongly Agree (5) to Strongly Disagree (1).

To test our hypotheses, an empirical study was conducted. Data were collected through questionnaire. The respondents median age was between 25 and 40 years with female 52% and male 48% ($n = 150$) was participated in the survey. The researchers used purposive sampling method to choose the respondents. The hypotheses were formulated to test the relationship between sustainability and citizen satisfaction.

H_1: Sustainability is positively related to Citizen Satisfaction.

H_2: e-Government services adoption is positively results Citizen Satisfaction.

4.4 Conceptual Model

The researchers proposed two models for this study; the first model involves three variables such as Sustainability, e-Government Services Adoption and Citizen Satisfaction. The relationship between these variables was investigated through primary data. The second model encompasses of Continuous improvement Approach and Citizen Satisfaction. This was validated through secondary data and case study method. The researchers, in both the model, study the influence of Sustainability and Continuous improvement approach to achieve the higher level of citizen satisfaction. The model is illustrated in Figure 4.1.

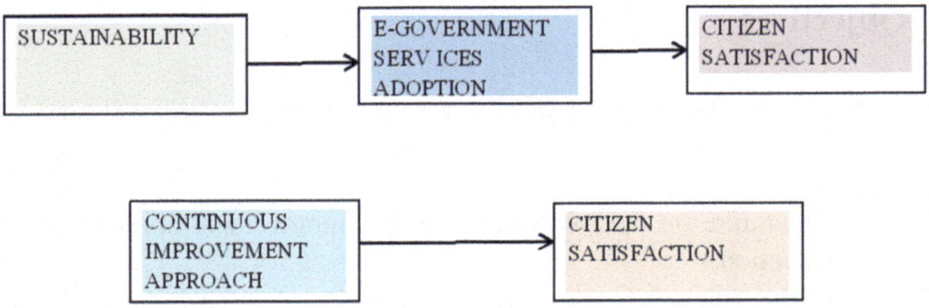

Fig. 4.1 Conceptual model of e-Government Services towards Citizen Satisfaction through Continuous Improvement Approach

4.5 Literature Review

4.5.1 Citizen Satisfaction

[4]. They carried out a number of trials to develop a useful instrument to evaluate the usefulness of various websites. This tool was first referred to as WebQual and then E-Qual. For continued use in bookstores, wireless news websites, and online auction sites, it was assessed and adjusted. The tool was used to assess the site design, usability, and content quality of three online auction sites [5].. Usability was used instead of site design quality when the tool for online bookshops was being tested by the researchers [6]. The concept of "usability" refers to how a user engages with and utilises a website.

4.5.2 Sustainability

In their study they found that addition to the support of the populace, representative governments also require the willing participation of their public employees [7]. The result of politics, policies, and actions is governance. In contrast to government, governance prioritises long-term processes over short-term outcomes. Because of this emphasis on procedures rather than judgements, governance today places a greater emphasis on "goals" than on rules [5]. From a governance standpoint, the goal is more significant than the procedures for achieving it.

4.5.3 e-Government Adoption

In their 2003 study on e-Government, [8] focused on the UK government's website quality. The Inland Revenue maintains the webpage. A new system that enables the online submission of self-assessed tax returns became operational immediately after the poll's opening. A comprehensive framework for evaluating e-Government offerings is provided by the tool, known as E-Qual, which draws on earlier research in the fields of information quality, service interaction quality, and website usability. In-depth information about how users saw the online taxation system when they attempted to interact with it was provided by the measurements and qualitative comments.

4.5.4 Continuous Improvement Approach

They found that outcomes rather than government outputs are the "bottom line" for governance [9]. Simply expressed, the purpose of governance outcomes is to produce results, as opposed to the purpose of government outputs, which is to produce something. People frequently lose patience with governments for a variety of reasons, including the fact that despite reports of significant effort, the results are typically viewed as being unsatisfactory by the general public. The ambition of those who wish to "re-invent government" is that it will change from an output-focused strategy to one that focuses outcomes [10]. People in positions of leadership commonly make the mistake of fusing the concepts of sustainability with the continuous improvement strategy.

4.6 Continuous Improvement Approach—Process

The process of continuous improvement approach is explained with the basis of its five features as follows.

4.6.1 Accessibility

The term "Agile accessibility" describes how accessibility best practises and considerations are incorporated into the agile software development approach. When developing software, the agile process places a strong emphasis on iterative development, collaboration, and adaptability in order to satisfy the changing needs of customers and stakeholders. Agile development practises that incorporate accessibility make guarantee that digital goods are inclusive and usable for people with impairments right from the start [11].

Governments must redesign their systems to make them more accessible and customer-focused. Modernising public administrations' organisational structures, operational processes, and overall culture should be a top goal for e-Government efforts. The OECD is one organisation that has stressed the significance of perceiving e-Government as a catalyst for innovation and transformation as opposed to an automation technology that only 'e-facelifts' existing practise.

4.6.2 Streamlined Services

"Streamlined services" relate to operations, procedures, or products that have been streamlined or optimised for effectiveness, efficiency, and enhanced user experience. Eliminating pointless procedures, simplifying processes, and improving overall service quality are the objectives of service streamlining [12]. To produce better results and higher levels of customer satisfaction, this notion is frequently implemented across numerous industries and sectors.

The existing digital gap, on the other hand, is largely seen by the United Nations as a significant barrier to e-readiness and e-evolution. Since only the wealthy actually have access to online opportunities, this disparity is fundamentally a global problem that appears in both developed and developing nations. However, it is indisputable that when and where e-Government is prioritised [13].

4.6.3 State-of-the-Art Technology

This refers to the most innovative and cutting-edge technological products, devices, tools, equipment, or systems currently on the market. It embodies the maximum degree of sophistication and development in a particular subject and stands for the cutting edge of innovation [14]. Modern technology is distinguished by its capacity to go beyond what was previously possible and to offer meaningful enhancements in functionality, performance, or user experience. It is simple to use these online services. Due to digital communication, more people are engaging in significant activities including voting, applying for e-visas, and more. African countries are replacing their typically inefficient paper-based file systems with e-Government technologies in an effort to lower the cost of government spending on services. Therefore, in the long run, electronic government systems are both economical and environmentally responsible [15].

4.6.4 Awareness

The term "awareness" in the context of sustainability refers to a general comprehension and awareness of the significance of sustainable practises and their effects on the economy, society, and the environment [16]. Recognising the limited nature of resources, the necessity of reducing environmental deterioration, the importance of fostering social fairness and economic stability for current and future generations, and many other factors are all part of it.

As a result, the computerization of internal processes and services constituted the initial stage of e-Government for developing countries. However, e-Governance is not a magic bullet for cost reductions, efficient, ethical governance, or economic advancement. On the other hand, e-Governance is a developing process that is frequently challenging, risky, and expensive from a political and financial standpoint [17].

4.6.5 Usability

Usability is important for a variety of goods and services, such as software programmes, websites, mobile applications, consumer gadgets, business equipment, and more. Usability testing, user research, and iterative design are frequently used to improve usability by identifying and addressing problem areas. In the end, a highly useable product adds to customer loyalty, higher productivity, and better overall results in addition to improving the user experience [18].

e-Government initiatives that are poorly conceived and implemented run the risk of wasting money, failing to provide the valuable services promised, and raising public dissatisfaction with the government. e-Government in poor nations must also include a variety of particular requirements, expectations, and challenges [8]. Emerging countries may, for example, have poor infrastructure, dishonest cultural practises, subpar educational systems, and unequal access to technology.

4.7 Enablers and Barriers to e-Government Success

e-Government provides citizen-centric, collaborative, "seamless," polycentric modern governance and ICT-based public administration modernization [19]. This comprehensive definition encompasses the qualities listed below, which sum up the roles of the government as a whole:

1. Government information programmes and services are delivered electronically through e-services.
2. E-commerce is the electronic exchange of money for products and services, such as when people pay their taxes and utility bills, renew their vehicle registrations, or spend money on leisure activities. It also includes the government's purchases of office supplies and surplus equipment.
3. Applying ICTs to better manage government, from streamlining administrative procedures to enhancing information exchange between government departments, is known as e-management.

4.8 Continuous Improvement Approach—Engineering Methodology

The general process of Continuous Improvement Approach comprises of five aspects which are explained in the below Fig. 4.2. This involves Process, Performance, Commitment, Strategy and Quality hierarchically.

a. **Process**

The steps involved in the process of continuous improvement approach are as follows:

- Create commitment and governance procedures.
- List the participants, such as users (people), government agents, and politicians.
- Identify and assess the mission and values of your organisation.
- Identify the e-Governance process's structural requirements and scope.
- Identify issues with governance, such as those with effectiveness, efficiency, and governance-centricity.

b. **Performance**

The following methodology is used for improving the performance of the continuous improvement approach:

- Establish performance metrics for governance-centricity, efficacy, and efficiency.
- Defining and outlining the e-Governance process' architectural layout. Obtain information on performance in relation to the chosen indicators.
- Establish objectives for enhancing indicators through measurement and analysis.
- Identify business and technological architectures for e-Governance, evaluate performance gaps in processes, and prioritise processes for redesign or improvement.

Fig. 4.2 Process of continuous improvement approach

c. **Commitment**

Create a strategy for projects involving process redesign and improvement; for each project, simulate and analyse the current process, develop a replacement, gather resources for replacement, and manage the changeover to the improved process.

d. **Strategy**

- Provide a report on the procedure.
- Provides updates on the performance and auditing process.
- Develop and integrate systems for monitoring and improvement.
- Report sharing and feedback collection.

e. **Quality**

A crucial component of Total Quality Management (TQM) and Lean Six Sigma techniques is quality in continuous improvement. Through the detection and elimination of flaws, the reduction of waste, and the optimisation of processes, these approaches seek to continuously improve the quality of products, processes, and services.

4.9 Analysis and Discussions

The primary data analysis was carried out through percentage analysis, Mean and Standard deviation, Correlation and Regression. The below tabulation is the scale used for the study measurement of Sustainability and e-Government over Citizen Satisfaction (Table 4.1).

4.9.1 Demographic Data

The total sample size is 150. Among them 72 are male and 78 are female. i.e., 48% are male and 52% are female. The age group of the respondents are described in the table below; 9.3% of the respondents are of the age group 51 and above, 25.3% of the respondents are between the age group of 41–50 years, 33.3% of the samples are between the age group of 31–40 years and 32% of the respondents are between 21–30 years (Table 4.2).

Among the 150 samples, 6.7% have completed higher secondary, 11.3% have completed Diploma/ITI, 27.3% have completed Undergraduate and 44.7% are Post Graduates.

Table 4.1 Scale used for study measurement

Concept measured	Scale items	Author
Sustainability	Accessibility	Parasuram et al. (2000)
	Streamlined services	Saxena (1999)
	Usability	Swanson (2003)
	Awareness	Baker (2000)
	State of the art technology	Barnes et al. (2003)
e-Government adoption	High Profile	Parasuraman et al. (2000)
	Status symbol	Heeks (2001)
	Efficiency enhancement	Saxena KBC and B.S. Sahay (2003)
	Travel ease	Yong (2003)
	Feedback enquiry	Swanson (2003)
Citizen satisfaction	Quality	Parasuraman et al. (2000)
	Personalized service	Saxena (2000)
	Quick services	Heeks (2000)
	Reliability	Parasuraman et al. (2000)
	Accuracy	Baker (2000)

Table 4.2 Demographic Data

		Frequency	%
Gender	Male	72	48
	Female	78	52
Age (Years)	21–30	48	32
	31–40	50	33.3
	41–50	38	25.3
	Above 50	14	9.3
Educational qualification	Higher Secondary	10	6.7
	Technical (Diploma/ITI)	17	11.3
	Undergraduate	41	27.3
	Post graduate	67	44.7
	Others	15	10

Table 4.3 Mean and standard deviation

Descriptive statistics			
	Mean	Std. deviation	Rank
e-Governance adoption	3.915	0.627	1
Citizen satisfaction	3.787	0.698	2
Sustainability	3.726	0.653	3

Table 4.4 Correlation

Correlations			
	Sustainability	e-Governance adoption	Citizen satisfaction
Sustainability	1	0.161*	0.144
e-Governance adoption		1	0.804**
Citizen satisfaction			1

*Correlation is significant at the 0.05 level (2-tailed)
**Correlation is significant at the 0.01 level (2-tailed)

4.9.2 Mean and Standard Deviation of Sustainability, Citizen Satisfaction and e-Governance Adoption

From the above descriptive statistics Table 4.3, it is evident that the highest mean score that the respondents believed is e-Governance Adoption (3.915), followed by Citizen Satisfaction (3.787) and then Sustainability (3.726).

4.9.3 Correlation Analysis

$\mathbf{H_{1(1)}}$: **There is significant** relationship between the sustainability, e-Governance adoption and citizen satisfaction (Table 4.4).

The researchers have done Pearson's Correlation analysis to examine the relationship between sustainability, e-Governance adoption and citizen satisfaction. From the above Table it is seen that all the variables are positively correlated and are significant at 1% level. The relation between sustainability and citizen satisfaction is 0.161 which denotes that there 16.1% of positive relationship between sustainability and citizen satisfaction at 1% significance level. The relationship between e-Governance adoption and citizen satisfaction is 0.804. The result indicates that there is 80% significant positive relationship between e-Governance adoption and citizen satisfaction. Hence the hypothesis $\mathbf{H_{1(1)}}$ is accepted.

4.9.4 Regression Analysis

$H_1$2: There is a significant influence of sustainability and e-Governance adoption on citizen satisfaction.

Independent variables: Sustainability; e-Governance Adoption.

Dependent Variable: Citizen Satisfaction.

The above Table 4.5 represents the model summary of the regression analysis. The value of R is 0.805.The relationship between the actual values and the predicted values of the citizen satisfaction is measured by R. It can be seen that e-Governance adoption and sustainability have 80% strong positive relationship with citizen satisfaction (Tables 4.6 and 4.7).

The predicted scores on Citizen Satisfaction are statistically significant in the model proposed for the study.

The coefficient of sustainability is 0.047 which is partial effect on citizen satisfaction and implies that citizen satisfaction would increase by 0.047 for each unit increase in sustainability and the effect is significant ($p < 0.05$). The e-Governance adoption coefficient is 0.797, which is the partial effect on citizen satisfaction keeping other variables constant meaning that citizen satisfaction would increase by 0.797 for each unit increase in e-Governance adoption and which is significant at 1% level.

Table 4.5 Regression analysis

Model	R	R square	Adjusted R square	Std. error of the estimate
1	0.805[a]	0.648	0.643	0.41735

Table 4.6 ANOVA

Model		Sum of squares	df	Mean Square	F	Sig
1	Regression	47.169	2	23.584	135.402	0.000[b]
	Residual	25.605	147	0.174		
	Total	72.773	149			

Table 4.7 Co-efficient

Model		Standardised		Unstandardised	t	P value
		B	Std. error	Beta		
1	(Constant)	0.126	0.272		0.461	0.645
	Sustainability	0.050	0.053	0.047	0.944	0.034
	e-Governance Adoption	0.888	0.055	0.797	16.122	0.000

[a]Dependent Variable: Citizen Satisfaction

Hence there is significant influence of sustainability and e-Governance adoption on citizen satisfaction is accepted ($H_1 2$).

4.10 Case Study Method

To establish the relationship between Continuous Improvement Approach and Citizen Satisfaction, various cases are analysed across the globe namely:

1. Europe
2. Asia
3. Africa
4. Australia
5. United States.

The dimensions of Continuous improvement approach such as User feedback, Service Delivery Digital Inclusion, Cost saving and process efficiency are assessed in depth in above countries through case study method. Despite widespread economic restructuring across many of them, including India, governmental administration in emerging countries remained highly bureaucratized and extremely centralised [19].

The availability of ICT infrastructure is a further differentiator between e-Governance in industrialised and underdeveloped nations. The development of accessible, remote access to governmental institutions made available by internet technology in industrialised countries was a major catalyst for the emergence of e-Governance [20]. However, government organisations were already implementing ICT-based solutions internally. However, poor countries had little to no ICT infrastructure, if any at all, because their public sector ICT usage was so low.

Status of e-Government Around the World

To assess the state of e-Government, very few indicators have been developed. This index assesses how e-Government services have evolved across UN member states. There are numerous things to consider. Additional measures for evaluating human capital include the proportion of urban to rural people. They conducted research, and as a result, they produced a fictitious statistic known as the "e-Government Index."

A second survey taken in 2003 redefined the e-Government Index in a different way, it is a composite index made up of the Web Measure Index, the Telecommunication Infrastructure Index, and the Human Capital Index. The E-Participation Index serves as a substitute index. Below is a description about these.

4.10.1 Europe

First, the European Union needs to guarantee that every one of its citizens has unrestricted access to Internet connectivity. Europe 2005 [21] also states that "member states should ensure that basic public services, where relevant, exploit the potential of multi-platform access"; to support these goals, education and training actions are considered essential to ensure that citizens will have the digital literacy and skills necessary to fully utilise the services. e-Government serves two purposes: it improves public administration's affordability and usability while also offering some data and value to back up the efforts made.

4.10.2 United States

This proposal contends that while e-Government is crucial for meeting the demands of business and citizen interaction with the Administration, it also coordinate their access to one-stop shops for services that will present a clear and concise image of the US government regardless of the access method (whether it be a phone call or a mouse click). All residents having access to the internet is no longer the main objective, but it is largely being accomplished. The OECD makes the following statements in order to prioritise and justify e-Government investments: Not all services can be made available online in a way that is attractive to consumers and companies [22].

4.10.3 Asia

Asian e-Government websites employ an approach that focuses dependability, linkage, content, usability, and self-service [23]. In order to combine pertinent content from both the connected site and other sites, a website must have a certain number and quality of links. Content is defined as information that is brief, relevant, and current in addition to how factual information is presented and organised online. Reliability refers to the technical performance of the website, including its availability and ability to operate as intended. Usability refers to how easily users can navigate a website and aims to lessen customer annoyance. On the other side, self-service techniques let clients do tasks quickly and easily.

The process of tailoring websites to the tastes or traits of specific users is related to the customization criterion. The phrase "information/content quality" refers to the calibre of the information systems' material and serves as a gauge for the advantages that the information offers to its users. The science and art of website navigation is what enables visitors to move between pages. The ability of a system to be used by as many individuals as feasible without change is referred to as accessibility. Providing citizens with technical support over email, chat, voice, and the internet is the main goal of web customer service [24].

4.10.4 Africa

Unfortunately, there are still several obstacles that drastically limit how basic services are provided in African nations, including fragmented and inconsistent communication across governmental departments and nations. Ineffective government service coordination makes it difficult to access information and important records, which, in turn, hinders effective governance. Additionally, the problems are exacerbated by the continent of Africa's expanding urbanisation and population. African governments, on the other hand, are working to improve the effectiveness and efficiency of their fundamental services. Their efforts are, however, constrained by efficient government operational procedures as well as by economic and budgetary imbalances. As a result, when essential services aren't provided, their inhabitants put pressure on the government to improve them and attack them when they don't.

Through the provision of e-Government services, African nations are gradually implementing digital and electronic government operating processes. Notably, ICT solutions, like digital technology and web-based apps, are being used by e-Government systems to improve consumer and business access to the provision of essential services across all governmental departments. For instance, in the 2018 report, numerous African nations were ranked as having a high level of e-Government development, including Ghana, Mauritius, South Africa, and Tunisia. This is a result of different online public services being offered in diverse African nations [25].

4.10.5 Australia

Especially from the perspective of those who want to access and use government information and services, it is difficult to say whether the governance reforms implemented by the federal government over the previous 10 years have enhanced efficiency and effectiveness. The creation of the National Office of the Information Economy in 1997 made a significant contribution to the development of the legal and regulatory framework, in addition to providing leadership across a very broad and diverse range of functions. The Information Management plan Committee's job is to support an all-encompassing or whole-of-government plan.

The AGIMO considers itself to be assisting the IMSC in fulfilling its mission by "identifying and advancing solutions to whole-of-government issues." Progress was being made, as seen by improved agency coordination and the development of more integrated services, according to AGIMO's most recent annual report. It discussed, for instance, the publication of a draughts paper for the Australian Government Authentication Framework, the Technology Interoperability Framework, and the introduction of the Fed Link encryption system (www.fedlink.gov.au) to increase the security of communications between agencies. In its Information Economy Index and Current State of Play reports, DCITA offers data on topics more pertinent to the digital divide, such as Internet access and usage [26].

A pilot project involving Centre Link and local governments in Western Australia and Queensland is one of the cross-jurisdictional projects being carried out under the National Service Improvement Framework. By delivering real-time eligibility information, this project gives participating local government access to Centre link's Customer Confirmation Service, enabling them to serve their clients more efficiently and effectively. A number of pilot programmes were implemented, including an online student bus pass application system with online eligibility validation provided by Centre link and Fish online, a service for recreational fishermen (https://eform.dier.tas.gov.au/sbpoaf/).

e-Governance—Comparative Views

The way the government uses technology, especially web-based Internet applications, affects how well citizens may access and receive government services and information as well as how well such services are provided to other government agencies, business partners, and staff members [15]. As a result, the techno-centric point of view actually gives a lot of weight to technological capabilities and the efficiency that they produce.

While an online-based automated government does work, the opposite isn't always true. Absent a democratic and open foundation, automation in the public sector won't always lead to better or more transparent government [16]. When employed to accomplish pertinent and focused goals, ICTs have the potential to revolutionise governance (Table 4.8).

The ability of ICTs to foster active, non-hierarchical information interchange is the cornerstone of their promise for transformation. In reality, if ICTs are used inappropriately, they can be used as a tool to undermine democracy and civic engagement. An alternate perspective to techno-centricity is "government-centric" (or citizen-centric) e-Governance. Such an approach to e-Governance places more attention on the impact or result of the method than just its outcomes [17]. As a result, a governance-centric point of view can assume that the reason why some government services are of lower quality than others is because they are either general (such as tax collection) or restricted to specific groups of people (such as the unemployed).

Table 4.8 Evaluation and comparison of Continuous Improvement Approach performance index

	Europe	United states	Asia	Africa	Australia
Streamlined Services	High	High	High	Medium	High
Agile Accessibility	High	Medium	High	High	Medium
Usability	High	High	Medium	Medium	High
Awareness	High	High	High	Medium	High
State of the art Technology	High	Medium	Medium	Low	Medium
Citizen Satisfaction	High	High	High	Medium	High
User feedback	High	Medium	Medium	High	Medium

4.10.6 e-Governance—a Methodology for Excellence in e-Governance

Customers, or citizens, are at the centre of companies that are process-focused, purpose-driven, and supported by structures. This is somewhat reinforced by the definition of "e-Government" used by the Ministry of Labour and Government Administration (1999), which emphasises the achievement of the following goals:

- Efficient and goal-oriented management—Government management must, to the maximum extent feasible, achieve "results" in accordance with set objectives, and these objectives must be met without the superfluous use of resources. Therefore, achieving excellence requires focusing on objectives or being driven by a purpose.
- Provide user-centered (citizen-centered) management—Governmental administration should acquire used to the preferences of its users (people) and, whenever possible, accept its operational procedures to suit their preferences.
- Open and democratic governance—requires that government operations be conducted in a way that promotes predictability, fair treatment, and a focus on transparency and the right to access information and make decisions.

This statement confirms once more that a purpose must be both purpose-driven and governance-centric in order to attain excellence [34]. Leadership, that is popular politically Government administration is a versatile and adaptable tool for carrying out governmental policies. What we mean by "governance-centric" e-Governance in essence is this.

4.10.7 Developing a Process-Oriented View of Government Work

The normal way that government work is done is via vertical, rigid "silos" of departments (or agencies), each of which does its tasks independently of the others.

- Setting up a performance management system that evaluates and tracks service performance on a regular basis in order to provide services effectively and efficiently. Due to the fact that it also places a high priority on service efficacy, this measuring approach is consistent.
- Building a scalable, adaptable, and cost-effective IT infrastructure for the government's high-volume operations. Additionally, users should find it secure and simple to use [35].

Due to their rigidity and/or deficiencies in one or more of these areas, many vendor-driven e-Governance solutions prove to be unsuitable in the long run. As a result, rather than simply computerising government processes, "excellent e-Governance" demands reform.

One strategy to ensure this integration is to take a rigorous approach to organising, creating, and deploying e-Governance services and technologies. Developing methodological frameworks for e-Governance has been attempted. The process also attempts to assist in creating a technology architecture that is compatible with the functioning and architecture of business processes [27].

One such approach is being developed by the current author and is known as "e-Governance engineering" (EGE). To accomplish the required integration, the approach consists of the following stages, with the following actions occurring at each stage [36].

4.10.7.1 Quality Assessment

The graph below depicts the four levels of quality assessment. Although this is the case, Fig. 4.2 shows that although the layers for the site interface and customer satisfaction emphasise citizen assessment, the layers for the process performance and technical performance emphasise organisation self-evaluation. The literature quality techniques that have been evaluated contain a large number of quality features and aspects [28]. The first axis for classifying the approaches under examination is based on the four levels that have been identified. The classification process includes two more axes: true/alternative and objective/subjective (Fig. 4.3).

The entire development process is the main attribute of this methodology. Second, an architectural base and a performance analysis of the complete e-Governance process are designed. As a result, there is a very low likelihood that the process will fail to fulfil its performance requirements once it has been put into place. Finally, to ensure the objectives, its performance is simultaneously audited and monitored [37].

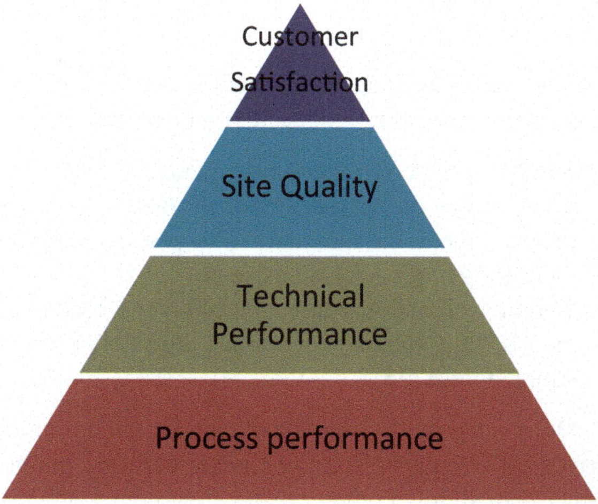

Fig. 4.3 Layers of quality assessment

- Interacting with residents
- Getting feedback from locals and
- Improving public services

4.11 The Web Measure Index

The Web Measure Index for 2005 is based on the e-Government five-stage concept.

- **Information in Stage I's Emerging Presence** is incomplete and basic. e-Government's online presence consists of links to the ministries/departments of education, labour, and ties to regional/local government are optional [18]. The majority of information is static with the fewest dynamic parts; nevertheless, some archival information.

S. No	Country	Index
1	United States	0.9062
2	Europe	0.8777
3	Australia	0.8679
4	Africa	0.7927
5	North America	0.8744
6	Asia	0.8727

- **In Stage II,** the government provides more up-to-date and archival information sources on governance and public policy, such as reports, newsletters, policies, laws, and regulations. The user can perform a document search in addition to using the site map and the help function. Information continues to primarily flow from the government to the citizen, even when the communication is more complex.
- **Interactive Presence is Stage III,** where interactive features are added to online government services to promote user convenience, such as the ability to download forms for paying taxes and applying for licence renewals. There is a capacity for both audio and video for important public information. You can get in touch with the government representatives by phone, fax, letter, and email. The website is updated more frequently in order to keep the content current and relevant for the general public.
- **Stage IV's Transactional Presence** makes it possible for citizens to have two-way communication with their government. It provides 24/7 online options for tax payments, ID card requests, birth certificate requests, licence renewals, and other C2G operations [29]. With their credit, bank, or debit card, citizens can pay for necessary government services like fines for transportation violations, taxes, and postal service costs. Suppliers of goods and services may submit bids for public interactions online through secure channels.

- **The fifth stage or Networked Presence** is the highest level of e-Government. The administration encourages public participation and is eager and equipped to engage the public [30]. The use of interactive elements and cutting-edge tools, allows soliciting citizen comments on legislation. The entity implies a shared understanding of democratic ideas like citizen empowerment and participatory decision-making.

4.11.1 Telecommunications Infrastructure Index

Six important ICT infrastructure capacity indicators are taken into account by the weighted average composite index known as the Telecommunication Infrastructure Index. The ratio of total gross enrolment for elementary, secondary, and higher education, as well as the adult literacy rate, make up the UNDP's "education index," which is used to calculate the Human Capital Index [31]. Adult literacy is assigned a value of two thirds, whereas gross enrolment is weighted at one third.

4.11.2 e-Participation Index

The government of a nation's information and services are assessed for their worth and quality in an effort to engage the general public in concerns of public policy [19]. The government's capacity and desire to inspire its citizens are gauged by this indicator.

4.11.3 e-Government Around the World

In terms of the nations, the USA comes in first (0.9062), followed by Denmark (0.9058). The United Kingdom (0.8777) has been passed by Sweden (0.8983), who is now in third place worldwide [32]. The highest emerging country is the Republic of Korea (0.8727), which is followed by Singapore (0.8503).

- An approach to version control.
- A continuous integration system.
- An integrated automation procedure.

4.12 Findings and Suggestions

The results suggest that Sustainability has positive influence over Citizen Satisfaction, i.e., as sustainability increases, the level of citizen satisfaction also increases. The second hypothesis is also proved which states, that e-Government adoption has a positive influence over citizen satisfaction. Through the case study method, the researchers have come to a conclusion that the Citizen satisfaction is highly positively influenced by Continuous Improvement Approach. The authors suggest the governments to carry out innovative continuous improvement methods to satisfy the citizen using e-Government services.

4.13 Policy Implications

A deliberate effort is made after CI programmes have been institutionalised successfully inside the organisations. Only the continuing deployment of excellent CI projects can enable manufacturing advances [33]. After the successful implementation of CI, the organization's current objective is to carry the CI Programme into the incremental process improvement phase utilising a CQI method. An organisation must keep on pushing forward after winning the CI excellence award in order to sustain its current levels of achievement and reach even higher ones [34]. The CI actions that have changed the organisation must be anchored so that they become ingrained in everyone's everyday routine. Driving Values of CI Respect, accountability, and compassion, the goals of CI Customer Contentment Enabling Values for CI The organization's humility, trust, openness, and cooperation may diminish once the project is finished.

Once the significant achievements made possible by strategically designed and institutionalised CI projects realised, consistent efforts should be made to identify new opportunities to build on their achievements [20]. As fundamental organisational values, transparency, trust, teamwork, continuous progress, and learning must always be communicated to staff [21]. Positive attitudes must permeate the entire organisation if CI is to be sustained (Fig. 4.4).

Similar to how CI may be used in conjunction with other important lean manufacturing initiatives like JIT, TQM, QFD, and TEI to improve overall manufacturing operations performances; this may also be used to examine production performance. Therefore, continuing KAIZEN activities have the potential to assist organisations in achieving their goals [22].

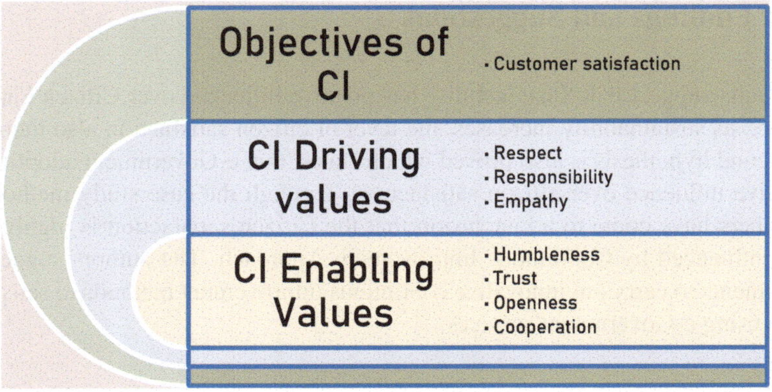

Fig. 4.4 Futuristic approach of continuous improvement

4.14 Conclusion

Projects involving e-Governance in numerous nations have failed to live up to their promises. In order to create high-quality e-Governance systems, an e-Governance engineering technique has been provided in this work.

To enable practical, inexpensive, and cost-effective control of the crucial quality and performance components of the delivery, the architecture and methodologies described in this article were purposely kept simple. In this way, service delivery workflows are evaluated based on their ultimate results as opposed to their intermediate phase as part of the entire approach to quality and performance assessment. The proposed framework excludes all other parties and only covers the perspectives of the two major groups, namely end-users and service providers. This contrasts with more sophisticated methods of performance evaluation like the Value Measurement Methodology or balanced scorecard-based techniques.

Overall, the methodology discussed in this paper is intended to elaborate the success of Continuous Improvement Approach in the e-Government services using the Sustainability procedures.

The literature suggests that there is a substantial body of work on the KAIZEN philosophy, which presents an extensive picture of past approaches and international research initiatives. In order to comprehend the significance of CI practises utilised in diverse sectors of application, researchers, CI professionals, and anyone interested in this method will find the available material beneficial. Principles, methodologies, tools, and strategies for CI implementation will be useful to practitioners.

Employee awareness of the many strategies included and the deployment of these strategies in various situations, however, is a crucial component in the successful application of the CI approach. Since these awareness-related characteristics are essential throughout the majority of global sectors, more research is required to improve them. The applicability of the techniques arisen in recent years to various organisations must also be determined through research.

By using a continuous improvement strategy, e-Governance initiatives may adapt to changing needs, technological advancements, and user expectations, resulting in more effective and citizen-focused government services.

References

1. Kelly, B., & Vidgen, R. (2005). A quality framework for web site quality: user satisfaction and quality assurance. In *Special Interest Tracks and Posters of the 14th International Conference on World Wide Web. 2005.*
2. Barnes, S. J., & Vidgen, R. (2003). Measuring web site quality improvements: A case study of the forum on strategic management knowledge exchange. *Industrial Management & Data Systems, 103*(5), 297–309.
3. Barnes, S. J., & Vidgen, R. T. (2002). An integrative approach to the assessment of e-commerce quality. *Journal of Electronic Commerce Research, 3*(3), 114–127.
4. Cheng, B.-S., Jiang, D.-Y., & Riley, J. H. (2003). Organizational commitment, supervisory commitment, and employee outcomes in the Chinese context: Proximal hypothesis or global hypothesis? *Journal of Organizational Behavior: The International Journal of Industrial, Occupational and Organizational Psychology and Behavior, 24*(3), 313–334.
5. Kettl, D. F. (2002). Managing indirect government. In *The tools of government: A guide to the new governance* (p. 490510).
6. Barnes, S. J., & Richard, T. V. (2003). Assessing the quality of a cross-national e-Government web site: A study of the forum on strategic management knowledge exchange. In *36th Annual Hawaii International Conference on System Sciences, 2003. Proceedings of the IEEE.*
7. Von Haldenwang, C. (2004). Electronic government (e-Government) and development. *The European Journal of Development Research.* Retrieved 7 January 2004.
8. Pathak, D., Singh, G., Belwal, R., Smith, R. F. I. (2007). E-governance and corruption developments and issues in Ethiopia. *Public Organization Review.*
9. Kannan, V. R., & Tan, K. C. (2005). Just in time, total quality management, and supply chain management: Understanding their linkages and impact on business performance. *Omega, 33*(2), 153–162.
10. Moore, M. H. (1995). Recognizing public value
11. Parasuraman, A., Berry, L. L., & Zeithaml, V. A. (1991). Refinement and reassessment of the SERVQUAL scale. *Journal of Retailing, 67*(4), 420–450.
12. Chan, F. K. Y., Thong, J. Y. L., Brown, S. A., Venkatesh, V. (2020). Service design and citizen satisfaction with E-government services: A multidimensional perspective. *Public Administration Review*
13. "Osborne, D., & Gaebler, T. (1992). Schneider, ingrid. In *Reinventing government* (pp. 302–304). New York; (1995).
14. Heeks, R. (2001) Understanding e-governance for development.
15. Leitner, C. (2003). EGovernment in Europe: The state of affairs. *EPIAScope, 2003*(3), 37–39.
16. Saxena, K. B. C., & Sahay, B. S. (2000). Managing IT for world-class manufacturing: The Indian scenario. *International Journal of Information Management, 20*(1), 29–57.
17. Wood-Harper, T., et al. (1990). Information systems development for developing countries. In S.C. Bhatnagar, &, N. Bjørn-Andersen (Eds.) *Information technology in developing countries* (pp 23–39).
18. Yong, J. S. L. (2003) e-Government in Asia: Enabling public service innovation in the 21st century.
19. Laffan, B. (2002). The European commission: Promoting EU governance. In *Participatory governance: Political and societal implications* (pp. 121–139). VS Verlag für Sozialwissenschaften.

20. Sukasame, N. (2004). The development of e-service in Thai government. *BU Academic Review, 3*(1), 17–24.
21. Stowers, G. N. L. (2008). Issues in e-commerce and e-Government service delivery. In *Electronic government: Concepts, methodologies, tools, and applications. Igi Global* (pp 3900–3915).
22. Singh, J., & Singh, H. (2012). Continuous improvement approach: State-of-art review and future implications. *International Journal of Lean Six Sigma, 3*(2), 88–111.
23. Brown, D. (2005). Electronic government and public administration. *International Review of Administrative Sciences, 71*(2), 241–254.
24. Halaris, C., et al. (2007). Classification and synthesis of quality approaches in e-government services. *Internet Research, 17*(4), 378–401.
25. Gouscos, D., et al. (2007). A general model of performance and quality for one-stop e-Government service offerings. *Government Information Quarterly, 24*(4), 860–885.
26. Mfoihaya, S. A., & Yusof, M. M. (2015). e-Government project enquiry framework for a continuous improvement process, status in Malaysia and Comoros. In *Information science and applications*. Springer.
27. Saxena, K. B. C. (2005). Towards excellence in e-Governance. *International Journal of Public Sector Management, 18*(6), 498–513.
28. Palvia, S. C. J., & Sharma, S. S. (2007). e-Government and e-governance: definitions/domain framework and status around the world. In *International Conference on E-governance* (Vol. 5. No. 1).
29. Chan, F. K. Y., Thong, J. Y. L., Brown, S. A., & Venkatesh, V. (2021). Service design and citizen satisfaction with e-Government services: A multidimensional perspective. *Public Administration Review, 81*, 874–894. https://doi.org/10.1111/puar.13308
30. Ma, L., & Zheng, Y. (2019). National e-Government performance and citizen satisfaction: A multilevel analysis across European countries. *International Review of Administrative Sciences, 85*(3), 506–526. https://doi.org/10.1177/0020852317703691
31. Alkraiji, A. I. (2020). Citizen satisfaction with mandatory e-Government services: A conceptual framework and an empirical validation. *IEEE Access, 8*, 117253–117265. https://doi.org/10.1109/ACCESS.2020.3004541
32. Alkraiji, A. I. (2021). An examination of citizen satisfaction with mandatory e-Government services: Comparison of two information systems success models. *Transforming Government: People, Process and Policy, 15*(1), 36–58. https://doi.org/10.1108/TG-01-2020-0015
33. Gupta, K. P., Singh, S., & Bhaskar, P. (2016). Citizen adoption of e-Government: a literature review and conceptual framework. *Electronic Government, an International Journal, 12*(2), 160–185.
34. Sigwejo, A., & Pather, S. (2016). A citizen-centric framework for assessing e-Government effectiveness. *The Electronic Journal of Information Systems in Developing Countries, 74*, 1–27. https://doi.org/10.1002/j.1681-4835.2016.tb00542.x
35. Ghareeb, A. M., Darwish, N. R., & Hefney H. A. (2019). e-Government adoption: Literature review and a proposed citizen-centric model. *Electronic Government, an International Journal, 15*(4), 392–416.
36. Sachan, A., Kumar, R., & Kumar, R. (2018). Examining the impact of e-Government service process on user satisfaction. *Journal of Global Operations and Strategic Sourcing, 11*(3), 321–336. https://doi.org/10.1108/JGOSS-11-2017-0048
37. Vimala, D., Vasantha, S., Shanmathi, A. (2023). Conceptual model and data algorithm for modernization of e-Governance towards sustainable e-Government services. In C. Gaie, & M. Mehta (Eds.) *Recent advances in data and algorithms for e-Government. Artificial intelligence-enhanced software and systems engineering* (Vol 5). Springer. https://doi.org/10.1007/978-3-031-22408-9_1

Chapter 5
Usage of Modern API for Automization of Government Procedures

Rahul Arvindkumar Vaghela, Kamini Solanki, Rimpal R. Popat, Ila Rahul Vaghela, and Nidhi Chhangani

Abstract In today's digital age, governments worldwide are recognizing the importance of modernizing their processes to keep up with the demands of their citizens. One way to achieve this is through the implementation of modern API-based architectures that allow for the automation of government procedures. This chapter provides a comprehensive guide to modern API-based architectures for the automation of government procedures. The chapter detailed here covers all relevant topics, such as the basics of API design, the advantages of using APIs in government, and how to integrate APIs into existing government systems. The chapter further covers practical examples of how APIs can be used to automate government procedures, such as online payment systems, permit applications, and citizen feedback mechanisms. Additionally, the chapter explores the security and privacy concerns associated with the use of APIs in government and provides strategies for mitigating these risks. The chapter is going to cover the tools and strategies needed to implement them successfully. Summing up, through this chapter, we are elaborating on the effort for

Present Address:
R. A. Vaghela (✉) · R. R. Popat
ASOIT, Silver Oak University, Gota Ahmedabad, Ahmedabad, India
e-mail: rahulvaghela.it@socet.edu.in; ravaghela83@gmail.com

R. R. Popat
e-mail: popatrimpal.2711@gmail.com

K. Solanki
CMPICA, Charusat, Changa, India
e-mail: kaminisolanki.mca@charusat.ac.in

R. R. Popat
Computer Engineering Department, Silver Oak University, Ahmedabad, India

I. R. Vaghela
PD Engineer, E-Infochips Pvt Ltd. Ahmedabad, Ahmedabad, India
e-mail: rahulvaghela.it@socet.edu.in

N. Chhangani
SOCCA, Silver Oak University, Gota Ahmedabad, Ahmedabad, India
e-mail: nidhichhangani.ce@socet.edu.in

C. Gaie and M. Mehta (eds.), *Transforming Public Services—Combining Data and Algorithms to Fulfil Citizen's Expectations*, Intelligent Systems Reference Library 252, https://doi.org/10.1007/978-3-031-55575-6_5

the adoption of modern API-based architectures in governments worldwide with the aim of improving efficiency, transparency, and accessibility of government services.

Keywords Government automation · Modern API integration · Digital transformation · Government procedures · API-driven governance · Automated workflow · Public sector innovation · e-Government solution · API-driven services · Process automation · Governance efficiency · Digital governance

5.1 Introduction

In today's rapidly evolving digital landscape, governments worldwide are constantly striving to enhance their operations and better serve their citizens. The traditional, manual-based approaches to government procedures have become increasingly inefficient and time-consuming, leading to frustration among citizens and hindered productivity within government agencies. To address these challenges, governments are embracing modern technologies and exploring innovative solutions to automate and streamline their processes. One such solution gaining significant traction is the use of modern API-based architectures, which offer a transformative approach to government automation [1].

Application Programming Interface (API) refers to a set of protocols, tools, and definitions that enable different software applications to communicate and interact with each other. APIs act as intermediaries, facilitating the exchange of data and functionalities between various systems and applications.

Historically, APIs have played a crucial role in enabling integration between different software components within private enterprises. However, their potential in the public sector has only recently been recognized, and governments are increasingly leveraging APIs to enhance the efficiency, transparency, and accessibility of their services.

The adoption of modern API-based architectures in government procedures holds immense promise for revolutionizing the way public services are delivered. By integrating APIs into their systems, governments can automate a wide range of processes, eliminating manual intervention, reducing paperwork, and expediting the delivery of services [2]. This automation not only enhances efficiency but also ensures consistency and accuracy in handling government procedures. For instance, by utilizing APIs, citizens can access government services online, submit applications, track their progress, and receive updates in real-time, all from the comfort of their homes. This level of automation and digitization not only simplifies the experience for citizens but also enables government agencies to allocate their resources more effectively, freeing up valuable time and personnel for other critical tasks.

Moreover, modern API-based architectures promote seamless data integration and sharing between different government agencies. Traditionally, government data has been siloed, with each department or agency maintaining its separate databases and systems. This fragmentation often leads to duplication of efforts, redundant data entry, and inefficient communication between departments. However, by implementing APIs, governments can establish secure and standardized channels for data exchange, enabling a holistic view of citizen information and fostering interdepartmental collaboration. For example, when a citizen updates their personal information in one government system, the API can automatically propagate the changes to other relevant systems, ensuring data consistency and eliminating the need for citizens to provide the same information multiple times [3].

Furthermore, modern API-based architectures enable governments to leverage the power of third-party innovation and collaboration. By exposing their APIs to external developers, governments can encourage the creation of innovative applications and services that enhance citizen experiences. This approach, commonly referred to as "open APIs" or "API ecosystems," allows developers to build upon existing government services, creating new applications that integrate seamlessly with government systems. Through such collaboration, governments can tap into the collective expertise and creativity of the developer community, driving innovation and expanding the range of services available to citizens.

However, the adoption of modern API-based architectures in government procedures does come with its own set of challenges. One of the primary challenges is ensuring the security and privacy of sensitive citizen data. Government agencies hold vast amounts of personal and confidential information, and any vulnerabilities in API implementation could expose this data to unauthorized access or misuse. Therefore, robust security measures, including authentication, encryption, and access control mechanisms, must be implemented to safeguard citizen data throughout its lifecycle [4].

Additionally, the successful implementation of API-based architectures requires comprehensive planning, coordination, and collaboration among government agencies. Integration with legacy systems, standardization of data formats, and adherence to interoperability standards are critical aspects that need to be carefully addressed.

Furthermore, training and capacity-building programs should be put in place to familiarize government personnel with API technologies and ensure their effective utilization.

5.2 Designing APIs for Government Services: A Comprehensive Guide

Government agencies are increasingly adopting APIs to provide efficient and accessible services to citizens. APIs serve as a bridge between government systems and external developers, enabling the creation of innovative applications and platforms that leverage government data and functionalities. However, designing APIs for government services requires careful planning and adherence to certain principles to ensure usability, security, and scalability. This chapter outlines essential considerations and best practices for basic API design in the context of government services.

5.2.1 Simplicity and Consistency

Government APIs should prioritize simplicity and consistency to foster ease of use and developer adoption. Designers should aim for clear and concise API endpoints, well-defined request and response structures, and consistent naming conventions. By employing simplicity and maintaining a standardized approach across different services, government APIs become more intuitive and reduce the learning curve for developers.

5.2.2 Documentation and Developer Support

Comprehensive documentation is crucial for successful API integration. Government API documentation should include clear instructions, detailed endpoint descriptions, parameter specifications, and sample code. Furthermore, providing interactive tools such as sandbox environments or software development kits (SDKs) can facilitate rapid prototyping and experimentation. Offering developer support channels, such as forums or dedicated support teams, help address queries, troubleshoot issues, and gather feedback to improve the API's functionality.

5.2.3 Authentication and Authorization

Securing government APIs is paramount to protect sensitive data and ensure proper access controls. APIs should implement robust authentication mechanisms, such as OAuth or API keys, to verify the identity of users and restrict access to authorized individuals or applications. It is essential to follow industry-based practices for secure token management, token expiration policies, and granular authorization rules to safeguard against unauthorized access and potential misuse.

5.2.4 Data Privacy and Compliance

Government services often involve personal and sensitive data. API design must adhere to strict privacy regulations, such as data protection laws or privacy frameworks specific to the jurisdiction. This is crucial to anonymize or pseudonymize personal data when exposed via APIs, implement data encryption during transmission, and adopt secure storage practices. Conducting regular security audits and adhering to relevant compliance standards ensure that government APIs maintain the highest level of data privacy and integrity.

5.2.5 Versioning and Backward Compatibility

Government APIs may undergo updates and improvements over time. To minimize disruptions for existing users, versioning and backward compatibility practices should be employed. By maintaining a stable API version and communicating any breaking changes, developers can plan and adapt their applications accordingly. Employing versioning strategies, such as semantic versioning, allows for smooth transitions and prevents unexpected disruptions to services relying on the API.

5.2.6 Rate Limiting and Throttling

To manage resource allocation and prevent abuse or excessive usage, rate-limiting and throttling mechanisms should be implemented in government APIs. By setting limits on the number of requests per second or minute, APIs can prevent overloading and ensure fair usage among all consumers. Clear documentation of rate limits and appropriate error responses enable developers to handle rate limit violations effectively and build resilient applications.

5.2.7 Monitoring and Analytics

Monitoring and analytics capabilities are essential to track API usage, identify performance bottlenecks, and improve service quality. Implementing comprehensive logging, metrics, and error-tracking systems allows government agencies to monitor API performance, detect anomalies, and respond proactively to potential issues. Leveraging analytics insights helps optimize API design, anticipate capacity requirements, and identify areas for improvement [5].

5.3 Advantages of Using API Design in Government

This section provides an insight into how the government sector is utilizing and deploying the newly emerging modern API. It describes the significance of APIs within the realm of digital government. We delve into three distinct perspectives to shed light on the subject matter. Initially, we examine how government APIs can contribute to achieving government objectives. Afterwards, we explore the role of APIs in facilitating public service delivery. Last and foremost, we concentrate on the various roles a government can assume concerning APIs, with a particular emphasis on the role of being an API provider.

5.3.1 Data Accessibility and Transparency

APIs provide a standardized and structured way to expose government data and services to external entities, such as citizens, businesses, and developers. By implementing APIs, governments can make their data more accessible and transparent, empowering citizens to interact with government services, retrieve information, and build applications that leverage government data.

5.3.2 Service Integration and Interoperability

Government agencies often have multiple systems and databases that need to communicate and share data. APIs enable seamless integration and interoperability between disparate systems, allowing different government departments or agencies to exchange information in a standardized manner. This promotes collaboration, efficiency, and the ability to provide comprehensive services to citizens.

5.3.3 Developer Ecosystem and Innovation

APIs stimulate the creation of a developer ecosystem by encouraging external developers and businesses to leverage government data and services to build innovative applications. This can lead to the development of new tools, services, and solutions that enhance citizen experiences, solve civic problems, and drive economic growth. By fostering innovation through APIs, governments can harness the collective creativity and expertise of external developers.

5.3.4 Efficiency and Cost Reduction

APIs streamline processes by automating interactions between systems and enabling self-service capabilities for citizens. By exposing government services and data through APIs, citizens can access information and perform transactions digitally, reducing the need for manual, paper-based processes. This improves efficiency, reduces administrative costs, and enhances overall service delivery.

5.3.5 *Standardization and Reusability*

APIs promote standardization in government systems and services. By defining a consistent interface and data format through APIs, governments can ensure that different departments or agencies follow a common approach, making it easier to share and reuse components and services. This reduces redundancies, encourages collaboration, and enables the development of modular and scalable solutions.

5.3.6 *Security and Privacy*

APIs allow governments to implement robust security measures to protect data and ensure privacy. With proper authentication, authorization, and encryption mechanisms, APIs can safeguard sensitive information and control access to government services. By centralizing security controls through APIs, governments can enforce consistent security practices and monitor and audit API usage for compliance purposes.

5.3.7 *Data-Driven Decision Making*

APIs facilitate the collection and analysis of data related to government services. By capturing data on API usage, governments can gain insights into citizen behavior, service demand, and system performance. This data can inform evidence-based decision-making, helping governments optimize service delivery, identify areas for improvement, and allocate resources effectively.

Overall, API design in government empowers citizens, promotes collaboration, drives innovation, enhances efficiency, and improves the overall effectiveness of government services. It establishes a foundation for a more transparent, accessible, and citizen-centric government ecosystem [6].

5.4 Security and Privacy Concerns of API

The increasing incidence of data breaches worldwide in recent years can be attributed to three significant trends: connectivity, extensibility, and complexity. These three factors have played a crucial role in enabling and facilitating the occurrence of data breaches on a global scale.

APIs introduce unique security and privacy concerns that need to be carefully addressed. The following aspects should be taken into account.

5.4.1 Connectivity

APIs have emerged as a pivotal factor in driving connectivity. Their significant role in establishing seamless connections and enabling communication between different systems and applications has contributed to the advancement of data sharing and integration. APIs facilitate the exchange of information, services, and functionalities, allowing diverse platforms to interact and collaborate effectively. As a result, the widespread use of APIs has enhanced connectivity and created new opportunities for data breaches to occur [7].

In a digitally interconnected enterprise, the integration extends beyond modern, cutting-edge applications to include legacy systems as well. It is crucial to note that these legacy systems may lack support for the latest security protocols, including Transport Layer Security (TLS) for safeguarding data during transit. Additionally, the libraries utilized within these systems may harbor numerous widely recognized security vulnerabilities, which often remain unresolved due to the intricate challenges involved in upgrading to the most recent versions. Consequently, if a connected system is not meticulously planned and designed, it can swiftly transform into a breeding ground for security risks and vulnerabilities.

5.4.2 Extensibility

Many enterprise software solutions are currently being developed with a strong focus on extensibility. Emphasizing extensibility instead of modification has become a widely recognized design philosophy in the software industry. This approach involves creating software that can adapt to new requirements without the need to alter the existing source code. Instead, the software allows for the seamless integration of new components into the current system. Notably, both Google Chrome extensions and Firefox add-ons adhere to this concept. While most browser plugins and extensions are harmless, it is important to remain vigilant, as attackers can exploit them to install malware. If an attacker successfully tricks you into installing a malicious browser plugin, they can potentially bypass your browser's security measures, including TLS encryption, and gain access to sensitive credentials for popular websites like Facebook, Google, or Amazon. The risk extends beyond the installation of a single plugin; having multiple extensions installed in your browser increases the overall attack surface. Attackers may not need to create new plugins themselves, as they can leverage existing vulnerabilities in already-installed plugins to carry out their malicious activities [8].

5.4.3 Complexity

The complexity of source code and system design is a widely recognized factor contributing to security vulnerabilities. Research has shown that, beyond a certain point, the number of defects in an application increases exponentially relative to the number of lines of code. At the time of writing, Google's entire codebase for its Internet services comprised approximately 2 billion lines of code, while the Microsoft Windows operating system consisted of roughly 50 million lines of code. Consequently, as the volume of code increases, it becomes crucial to expand the number of tests conducted to ensure that existing functionalities remain intact and that newly added code operates as intended.

5.4.4 Authorization and Permissions

Implementing robust authorization mechanisms to control the actions and operations that API users can perform. Granular permissions should be enforced to restrict access to sensitive resources or functionalities.

5.4.5 API Abuse and Rate Limiting

Protecting against API abuse, such as denial-of-service attacks or excessive API requests, by implementing rate-limiting mechanisms. This helps prevent API overload and ensures fair resource allocation.

5.4.6 Secure Data Transmission

Ensuring secure transmission of data between the API client and server through secure communication protocols. This prevents the interception or tampering of sensitive information.

5.4.7 Logging and Monitoring

Implementing logging and monitoring mechanisms to track API usage, detect suspicious activities, and identify potential security breaches Real-time monitoring can help identify and respond to security incidents promptly.

5.4.8 Privacy Compliance

Adhering to privacy regulations and guidelines, such as the General Data Protection Regulation (GDPR), to protect user privacy APIs should handle personal data appropriately, including obtaining necessary consent and providing transparent privacy practices.

5.4.9 Secure Development Practices

Following secure coding practices and conducting regular security assessments, including vulnerability scanning and penetration testing, to identify and remediate potential security weaknesses in the API implementation.

Addressing these security and privacy concerns is crucial to ensuring the integrity, confidentiality, and availability of data transmitted through APIs, fostering trust among users, and protecting sensitive information from unauthorized access or misuse.

5.5 Cost and Challenges of API

The adoption of APIs comes with associated technical and organizational expenses. The extent of these costs can differ significantly depending on the government's specific role in API implementation. Furthermore, embracing APIs presents various challenges, including the need to overcome the required shift in organizational mindset, address security vulnerabilities, and ensure compliance with relevant regulations like the General Data Protection Regulation (GDPR). This section provides a concise summary of the qualitative analysis concerning the costs and challenges involved in API adoption.

From a technical standpoint, the costs associated with API adoption can vary significantly based on the extent of its implementation. These costs encompass a wide range of factors, including operational aspects like software development, deployment, and maintenance, as well as tactical considerations such as establishing digital infrastructures and enhancing capabilities. Additionally, strategic concerns come into play, involving the formulation of requirements aligned with the attainment of digital government objectives [9].

API adoption presents challenges at both the organizational and technical levels. At the organizational level, these challenges involve the need to foster a shift in organizational mindsets and address new barriers related to interoperability and organizational dynamics. To overcome these challenges, the study participants analyzed in the research have taken steps such as developing a shared vision of their digital ecosystem and implementing effective coordination efforts.

From a technical perspective, the primary concerns include heightened IT vulnerability, increased complexity in complying with existing regulations, and a shortage of skills in relevant areas. The evidence gathered from our study sheds light on these technical challenges, revealing the following trends and observations.

5.5.1 *Security*

The issue of cybersecurity is of paramount importance when dealing with APIs, as emphasized in the previous section. APIs serve as gateways for accessing an organization's network, exposing data, services, and transactions to facilitate the creation of new services. However, if these APIs are not adequately protected, they can become entry points for unauthorized access to the internal network, potentially leading to the exploitation of internal vulnerabilities. Thus, it is crucial to ensure the proper security measures are in place for APIs to safeguard data privacy and inspire confidence among citizens in the service delivery channel.

When APIs are designed to provide access to public data, they must also be safeguarded against inappropriate use or abuse, including denial of service attacks. Various security solutions exist, such as OAuth and certificate-based authentication, which can be implemented in conjunction with a comprehensive cybersecurity strategy and cryptographic measures.

5.5.2 *Adhering to Existing Regulation*

Externally facing public-sector APIs primarily deal with sensitive data, often containing information related to citizens. While these APIs can enhance government transparency, their adoption necessitates careful attention to regulatory compliance. Specifically, when publishing APIs and sharing data, adherence to various regulations is crucial.

5.5.3 Lack of Skills

To achieve successful implementation of API strategies in the public sector, it becomes necessary to acquire a new skill set and knowledge base. While policy stakeholders may not require in-depth technical knowledge of APIs, the transformative impact of APIs on government functioning calls for a broader understanding of their role in a digital government era. Leaders in particular will benefit from having a contextual understanding of APIs. Throughout the study, stakeholders consistently emphasized the need for improved storytelling surrounding the significance and value of APIs [10].

In a government that extensively adopts APIs, additional crucial skills come to the forefront. These include the capacity to evaluate the value derived from technology enablers, employ user-centered design practices, and possess ecosystem facilitation skills. As discussed earlier about funding models, the adoption of new collaborative work practices becomes necessary to dismantle entrenched departmental silos.

5.6 Integration of APIs into Existing Government Procedures

Integrating APIs into existing government procedures can lead to significant improvements in efficiency, transparency, and accessibility of services. APIs enable different software systems to communicate and exchange data seamlessly, making it possible for government agencies to streamline processes, reduce manual work, and enhance the overall citizen experience. Here is how the integration of APIs can be beneficial and the steps involved in the process (Fig. 5.1).

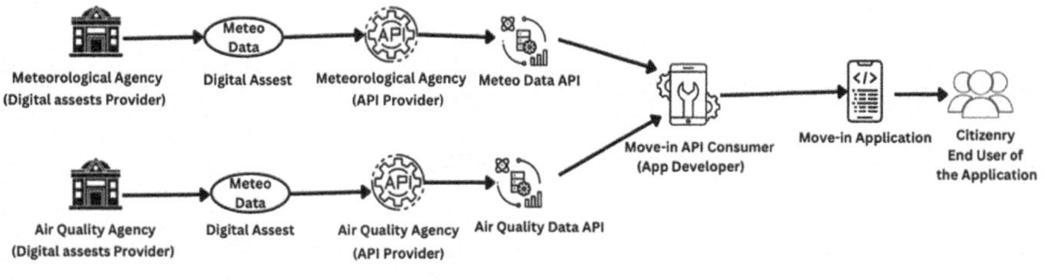

Fig. 5.1 Move-in API system [11]

5.6.1 Benefits of Integrating APIs into Government Procedures

5.6.1.1 Efficiency

APIs allow disparate government systems to share information in real time, reducing the need for manual data entry and minimizing the chances of errors. This streamlines processes and speeds up service delivery.

5.6.1.2 Transparency

Integrating APIs can lead to greater transparency as citizens can track the status of their applications or requests through online platforms. This helps build trust and confidence in government services.

5.6.1.3 Automation

APIs enable the automation of routine tasks, freeing up government employees to focus on more complex tasks that require human judgment. This can lead to increased productivity and improved service quality.

5.6.1.4 Data-Driven Decision Making

APIs provide access to valuable data that can be analyzed to make informed policy decisions and optimize government processes.

5.6.1.5 Accessibility

APIs facilitate the creation of user-friendly online platforms and mobile applications, making government services more accessible to citizens, even those who might face physical or geographic limitations.

5.6.2 Steps to Integrate APIs into Government Procedures

5.6.2.1 Identify Objectives

Determine the specific goals you want to achieve through API integration, such as enhancing service delivery, improving data sharing, or automating certain processes.

Assess Existing Systems

Evaluate the current government procedures and IT infrastructure to identify areas where APIs can be integrated effectively. Consider the compatibility of existing systems with modern API standards.

API Development

Develop or acquire the necessary APIs that will facilitate data exchange between different systems. Ensure that these APIs adhere to industry standards and security protocols.

Data Security and Privacy

Prioritize data security and privacy during the integration process. Implement strong authentication mechanisms and encryption protocols to protect sensitive citizen information.

5.6.2.2 Testing and Quality Assurance

Thoroughly test the APIs to ensure they function as intended and are capable of handling different scenarios. Conduct load testing to determine their performance under various levels of traffic.

5.6.2.3 Training

Provide training to government employees on how to use the newly integrated APIs. This ensures that staff members are comfortable with the technology and can utilize it effectively.

5.6.2.4 User Interface Design

Design user-friendly interfaces for the public to interact with government services through the integrated APIs. The user experience should be intuitive and accessible across different devices.

5.6.2.5 Feedback and Continuous Improvement

Encourage citizens and government employees to provide feedback on the newly integrated systems. Use this feedback to make necessary improvements and enhancements.

5.6.2.6 Monitoring and Maintenance

Regularly monitor the performance of the integrated APIs and address any issues promptly. Keep the APIs updated to stay in line with evolving technology and security standards.

5.6.2.7 Scale and Expand

Once successful integration is achieved in one area, consider expanding the use of APIs to other government procedures, departments, or agencies to reap the benefits across the board.

5.7 Tools and Strategies for Successful Implementation

The implementation of API-based architecture in government services has become increasingly important for delivering efficient and accessible digital solutions. APIs enable government agencies to unlock the potential of their data and services, empowering developers to create innovative applications and platforms. To ensure a successful implementation, government organizations need to adopt appropriate tools and strategies that facilitate the development, management, and security of APIs. This chapter explores key tools and strategies essential for the effective implementation of API-based architecture in government services [12].

5.7.1 API Management Platforms

API management platforms provide a comprehensive suite of tools and features to streamline the entire lifecycle of APIs. These platforms assist in designing, documenting, testing, deploying, and monitoring APIs. Government agencies can leverage API management platforms to create developer portals, interactive documentation, and sandbox environments for developers to explore and integrate with APIs. Popular API management platforms, such as Apigee, Mulesoft, or AWS API Gateway, offer extensive functionalities for API governance, security, analytics, and developer engagement.

5.7.2 Open Data Initiatives

Open data initiatives play a pivotal role in government API implementations, facilitating access to public information and promoting transparency. By providing structured and well-documented APIs for public datasets, government agencies empower developers and citizens to build applications that leverage government data. Tools like data catalogs and data marketplaces can help centralize and manage open data initiatives, ensuring the availability and discoverability of APIs that expose government information.

5.7.3 Interoperability Standards

Government services often involve multiple agencies and systems that need to communicate seamlessly. Interoperability standards, such as Representational State Transfer (REST) or Simple Object Access Protocol (SOAP), ensure compatibility and integration between different systems. Adhering to widely accepted interoperability standards enables efficient data exchange, promotes collaboration, and simplifies the integration of disparate government services.

5.7.4 Security and Access Control

Security is of paramount importance when implementing API-based architecture in government services. Robust security measures, such as authentication, authorization, and encryption, should be implemented to protect sensitive data and ensure authorized access. Tools like identity and access management (IAM) systems, OAuth, or OpenID Connect provide secure authentication and authorization mechanisms. Additionally, API gateways and firewalls can be employed to enforce security policies, inspect and filter API traffic, and protect against potential threats.

5.7.5 Developer Engagement and Support

Engaging with developers and providing adequate support is crucial for the success of government API implementations. Developer portals, forums, and communities can foster collaboration and encourage developers to create innovative applications using government APIs. Dedicated support channels, including documentation, code samples, and API usage policies, enable developers to understand and integrate with government services effectively. Regular developer outreach events, hackathons, and incentivization programs further promote engagement and adoption [13].

5.7.6 *Performance Monitoring and Scaling*

Monitoring and scaling tools are vital for ensuring the performance and scalability of government APIs. Monitoring tools, such as application performance monitoring (APM) solutions, help identify bottlenecks, track response times, and troubleshoot issues [14]. Load balancing and scaling techniques, including auto-scaling, containerization, and serverless architectures, enable government APIs to handle increased traffic and provide uninterrupted service during peak periods.

5.7.7 *Compliance and Governance*

Government agencies must adhere to regulatory compliance and governance requirements when implementing API-based architecture. Compliance tools and frameworks, such as the General Data Protection Regulation (GDPR) or the Federal Information Security Management Act (FISMA), ensure that data privacy and security standards are met. Implementing governance policies, such as API usage policies, versioning guidelines, and change management processes, fosters consistency, control, and accountability across government API implementations [15].

5.8 Cyber Threats in Government Systems: The Vulnerabilities Within API-Based Architectures

In the intricate landscape of government systems, the integration of Application Programming Interfaces (APIs) stands as a cornerstone for efficient data exchange and streamlined services. However, this technological advancement brings forth a labyrinth of cyber threats that pose significant risks to the integrity, confidentiality, and availability of critical government data. At the forefront of these threats lies the specter of API misuse and abuse. Malicious actors, leveraging vulnerabilities within APIs, exploit their functionalities by manipulating inputs or inundating systems with an influx of requests, a tactic known as API abuse. This exploitation, if left unchecked, results in severe repercussions, potentially compromising the entire system's stability and security.

Moreover, the perils of injection attacks loom large over API-based architectures within government systems. Like web applications, APIs are susceptible to infiltration by malicious code through SQL or command injections. Inadequate validation of inputs exposes vulnerabilities that adversaries exploit, facilitating unauthorized access to sensitive data or compromising the system's integrity. The insidious nature of cyber threats extends to the realm of insufficient authentication and authorization mechanisms. Weak authentication protocols or flawed authorization checks serve as gateways for unauthorized access. Malevolent entities capitalize on these weaknesses, infiltrating government networks and gaining illicit access to confidential information, undermining the sanctity of governmental operations. Data breaches stand as a grim reality in the face of inadequate encryption or mishandling of data transmitted through APIs. The repercussions of such breaches, especially when dealing with citizens' personal data or classified government information, have far-reaching consequences, eroding public trust and jeopardizing national security.

Furthermore, the specter of distributed denial of service (DDoS) attacks looms ominously over API-based architectures. These attacks, orchestrated to overwhelm systems with an avalanche of requests, render them inaccessible to legitimate users, thereby disrupting critical government services and plunging systems into chaos.

The absence or ineffectiveness of rate-limiting and throttling mechanisms within APIs exacerbates vulnerabilities. Without proper controls in place, the flood of excessive requests engulfs systems, resulting in instability or overload and creating a breeding ground for exploitation by cyber adversaries. Insecure integrations with third-party services or legacy systems stand as another Achilles' heel within API-based architectures. These integrations, if not fortified with stringent security measures, serve as gateways for attackers to exploit vulnerabilities and gain unwarranted access to the government network, perpetuating potential breaches and compromises. The crux of mitigating these multifaceted threats lies in robust security measures tailored explicitly for government systems. Implementing stringent authentication measures, such as OAuth, API keys, or token-based systems, coupled with robust access controls, forms the bedrock of defense against unauthorized access.

Moreover, encryption emerges as a formidable shield against cyber threats, safeguarding data in transit through the implementation of robust encryption protocols like TLS and SSL. Rigorous input validation protocols serve as an essential barricade against injection attacks, fortifying the system's resilience against malicious infiltration attempts.

Additionally, instituting stringent rate-limiting and throttling mechanisms acts as a bulwark against overload-induced vulnerabilities. Regular security audits, continuous monitoring, and timely updates serve as proactive measures in identifying and rectifying potential vulnerabilities before they escalate into full-blown security breaches.

In the dynamic landscape of government systems, securing API-based architectures demands an unwavering commitment to evolving security paradigms. The continuous adaptation to emerging threats, coupled with a proactive approach to fortifying vulnerabilities, is imperative to ensure the sanctity of government data and the unwavering trust of citizens in their governance.

5.9 Conclusion

In conclusion, the research on "Usage of Modern APIs for the Automation of Government Procedures" underscores the transformative impact of APIs in reshaping public administration. The integration of modern APIs streamlines government procedures, fostering efficiency, transparency, and citizen satisfaction. Through seamless data exchange and interoperability, APIs enable real-time communication between diverse systems, mitigating delays and enhancing service delivery. The collaborative potential of APIs, inviting external developers to contribute to government applications, promotes innovation and adaptability. However, it is crucial to address data security concerns to maintain public trust in this digital evolution. Overall, the study reveals that embracing modern APIs is pivotal for creating a responsive, citizen-centric government, marking a paradigm shift towards streamlined processes and improved public services.

References

1. Kharlamov, A., & Hartig, O. (2019). Government as a platform: A literature review of a proposed model for modern governments. *Government Information Quarterly, 36*(4), 691–706.
2. Dabrowski, M. (2021). Digital government services: Understanding the role of APIs in enabling seamless user experiences. In *Proceedings of the 22nd Annual International Conference on Digital Government Research* (pp. 13–24). ACM.
3. European Commission. (2020). API Strategy and Use in Government: Challenges and Opportunities. Retrieved from https://ec.europa.eu/isa2/sites/isa/files/isa2_api_stratgy_and_use_in_government_challenges_and_opportunities.pdf
4. Lee, G., & Kwak, Y. H. (2020). Assessing the success factors of API platforms in government services. *Sustainability, 12*(11), 4585.
5. Fang, Z., & Zhang, J. (2020). The development trend of government information system based on API technology. In *Proceedings of the 9th International Conference on Advanced Computer Science and Information Systems* (pp. 146–153). ACM.
6. Parycek, P., & Leitner, C. (2018). A common API for e-Government services: A technical blueprint. *International Journal of Public Administration in the Digital Age, 5*(3), 42–62.
7. World Bank. (2020). Maximizing the value of government APIs. Retrieved from https://www.worldbank.org/en/topic/digitalgovernment/brief/maximizing-the-value-of-government-apis
8. European Data Portal. (2021). API good practices for public administrations. Retrieved from https://www.europeandataportal.eu/sites/default/files/edp_landscaping_api_good_practices.pdf
9. W3C. (2021). Government linked data working group. Retrieved from https://www.w3.org/2011/gld/
10. U.S. General Services Administration. (2021). API Standards for U.S. Government. Retrieved from https://apievangelist.com/2016/10/20/api-standards-for-us-government/
11. United Nations Department of Economic and Social Affairs. (2020). The role of Application Programming Interfaces (APIs) for public services. Retrieved from https://publicadministration.un.org/en/Research/News-View/ArtMID/2284/ArticleID/4076/The-role-of-Application-Programming-Interfaces-APIs-for-Public-Services
12. Lacity, M. C., & Willcocks, L. P. (2020). Digital government at scale: Lessons learned. *Communications of the ACM, 63*(2), 22–25.

13. Miritello, G., Lara, R., & Moro, E. (2020). Exploring the usage of open data APIs for the development of mobile applications in smart cities. *Journal of Urban Technology, 27*(4), 19–39.
14. European Data Protection Supervisor. (2021). API Design Guidelines for the European Union. Retrieved from https://edps.europa.eu/sites/edp/files/publication/20-09-30_api_design_guidelines_for_the_eu_en.pdf
15. Geyer, C., & Brune, P. (2020). Public Administration in the API Economy. In *Proceedings of the 53rd Hawaii International Conference on System Sciences*.

Chapter 6
Epidemiology Inspired Cybersecurity Threats Forecasting Models Applied to e-Government

Jean Langlois-Berthelot, Christophe Gaie, and Jean-Fabrice Lebraty

Abstract This chapter delves into the innovative fusion of epidemiology and cybersecurity, presenting a novel paradigm for forecasting cyber threats with applications for e-Government. Drawing inspiration from epidemiological models that predict the spread of diseases, we propose pioneering approaches to anticipate and mitigate cyber threats in the digital governance landscape. To enhance the robustness of cyberattack forecasting, the chapter explores ensemble methods that combine predictions from multiple epidemiology models. This approach aims to mitigate individual model biases and improve forecasting accuracy. It also outlines that human expertise is required to contextualize the forecasts, identifying potential outliers, and define cybersecurity strategies. In conclusion, this chapter provides a comparison of the proposed models and identifies future challenges to enhance cybersecurity of e-Government.

Keywords Epidemiology · Forecasting · Cybersecurity · SEIR · Herd Immunity · e-Government · Algorithm · Digital Identity

J. Langlois-Berthelot
Senior Fellow, CEMS-T, French Ministry of Armed Forces, 75007 Paris, France
e-mail: jean.langlois@sciencespo.fr

C. Gaie
Chief of Engineering and Innovation Department, French Prime Minister Services, 75007 Paris, France

J.-F. Lebraty (✉)
Laboratoire de Recherche Magellan, Université Jean Moulin, 69372 Lyon, France
e-mail: jean-fabrice.lebraty@univ-lyon3.fr

Institut d'Administration Des Entreprises (IAE), Lyon, France

C. Gaie and M. Mehta (eds.), *Transforming Public Services—Combining Data and Algorithms to Fulfil Citizen's Expectations*, Intelligent Systems Reference Library 252, https://doi.org/10.1007/978-3-031-55575-6_6

6.1 Introduction

The evolution of e-Government concepts has provided valuable insights into the dynamic nature of governance in the digital age, highlighting the critical role of Information and Communication Technologies (ICT) in modernizing government processes. As governments around the world increasingly embrace digital transformation, the concepts of e-Government, e-Participation, and e-Governance have emerged to address the evolving landscape of public administration. While these concepts have significantly enhanced government services and citizen engagement, they also underscore the need for innovative approaches to forecast and mitigate cyber threats in this digital era.

e-Government, at its core, refers to the initial approach of leveraging ICT to modernize public services and streamline government procedures. It has evolved over the years to encompass a broader spectrum of research and development initiatives, including eParticipation, e-Governance, and even t-Government [1]. This evolution underscores the multidisciplinary nature of the field, which encompasses legal aspects, public administration, and computer science.

Grönlund and Horan's seminal work in 2005 provided a comprehensive review of e-Government [2], defining key concepts and emphasizing the importance of cross-domain research. They highlighted the necessity of transparent regulations to safeguard citizen rights and privacy while ensuring government efficiency. The importance. The importance of enhancing security of information was also outlined by Shareef who details both technical and non-technical security perspectives to enhance e-Government security [3].

E-Participation, a branch of e-Government, focuses on involving citizens in the design and delivery of government-related processes. It seeks to foster collaboration between civil servants and citizens, ultimately enhancing the participation of citizens in shaping public services. A detailed review of e-Participation was proposed by Santamaría-Philco and concerns the research from 2000 to 2019 [4]. The paper details the different dimensions involved in terms of process, actors, roles, etc. As e-Participation increases interactions between government and citizens, it also brings forth new challenges, particularly in the realm of cybersecurity.

The integration of ICT into government processes, as exemplified by e-Governance, presents a two-way protocol for governance, in contrast to the one-way approach of e-Government. E-Governance emphasizes the importance of citizen satisfaction and their active involvement in decision-making processes. While this evolution has improved government efficiency and transparency, it has also exposed governments to cybersecurity risks and challenges. Interestingly Singh and Karaulia outline various security issues that affect e-Governance [5].

In today's interconnected digital world, the need for new methods of forecasting cyber threats has never been more pressing. As government systems become increasingly digitized and interconnected, they become attractive targets for cybercriminals and state-sponsored hackers. These threats range from data breaches and identity theft to sophisticated cyberattacks that can disrupt critical infrastructure and compromise national security. Some pioneer researchers introduced the notion of forecasting in e-Government especially in the context of data availability [6]. To the authors' knowledge forecasting is not largely employed to enhance the security of government services.

The history of e-Government's evolution highlights the importance of addressing these cybersecurity challenges. For example, Reitz's work in 2006 emphasized the necessity of robust regulations to protect sensitive data used in e-Government initiatives [7]. Similarly, Alshehri and Drew (2010) identified various barriers to successful e-Government implementation, including security concerns [8]. These early insights underscored the importance of cybersecurity in the e-Government landscape.

In the context of e-Participation, cybersecurity becomes even more critical. Platforms that facilitate collaboration between citizens and government entities can be vulnerable to cyberattacks, potentially compromising sensitive citizen data and eroding trust in government. Ensuring the security of these platforms and the privacy of citizens participating in e-Government initiatives is paramount. To this end, Buccafuri et al. proposed a framework that enables to vote electronically while preserving privacy [9].

E-Governance, with its focus on citizen involvement and satisfaction, amplifies the need for robust cybersecurity measures. As governments increasingly rely on digital technologies to engage citizens and make informed decisions, they must also prioritize the protection of sensitive data and infrastructure. The transition from e-Government to e-Governance demands a proactive approach to cybersecurity to maintain public trust and ensure the integrity of democratic processes.

The evolution of e-Government concepts has illuminated the transformative power of ICT in government processes. However, this transformation has also exposed governments to new and evolving cyber threats. To navigate this digital landscape successfully, innovative methods of forecasting and mitigating cyber threats are essential. Cybersecurity must be at the forefront of e-Government, e-Participation, and e-Governance initiatives to safeguard citizen data, maintain public trust, and ensure the continued progress of governance in the digital age.

A very instructive review of IT security trends to fight against e-Government threats was proposed by Efe and Kazdal in [10]. First, the authors describe the potential attacks such as internal threats or malwares. Then, they outline the various trends to enhance the security of government services such as blockchain and data security, secure software development or tailor-made security. These measures increase the credibility of e-Government. Mishra et al. also outline the importance of public education to increase cyber safety of government webservices to ensure the confidence of citizens towards e-Government [11].

The landscape of digital governments, often referred to as e-Government, has evolved significantly over the years. Governments worldwide have embraced Information and Communication Technologies (ICT) to modernize public services and streamline administrative processes. However, this evolution has brought forth a host of challenges, particularly in the realm of cybersecurity. Fore these reasons, it is urgent to tackle the challenges faced by digital governments.

As the landscape of digital governments, often referred to as e-Government, has evolved significantly over the years with governments worldwide embracing Information and Communication Technologies (ICT) to modernize public services, it has become increasingly evident that this transformation comes with a significant challenge—the escalating threat of cyberattacks.

The integration of ICT into government processes, as exemplified by e-Government, has undoubtedly enhanced efficiency and accessibility. However, in the midst of this digital revolution, governments have also found themselves vulnerable to a wide array of cyber threats. These evolving threats necessitate not only advanced cybersecurity measures but also a proactive approach to anticipating and mitigating potential cyberattacks. To this end, there are multiple initiatives to implement privacy be design such as described by Gerunov in [12]. Moreover, in today's interconnected digital world, where malicious actors constantly seek to exploit vulnerabilities, the need for innovative and dynamic methods of forecasting cyber threats has become imperative.

In the current global landscape, cyber security has become a determinant issue for digital government with the growing risks of cyber-attacks. The fast and sustained evolution of cyber threats highlights the imperative need to understand and predict those attacks dynamically.

Over the past few decades, cyberspace has become the prime playground of a constant competition between malicious actors. The concept of cyber-attack refers to a large number of very different realities. These attacks strongly vary in size, sophistication and objective ranging from simple theft of personal data to devastating state cyberattacks. Business, critical infrastructures, government institutions and even individuals are exposed to these constant threats. Cyber-attacks take various forms like ransomware, phishing, distributed denial of service (DDoS) and many other tactics often with devastating financial and operational consequences. In a recent report, the FBI indicated that they received over 800,000 internet crime complaints, with an impact that exceeds $10.2 billion [13].

Additionally, the cyber threat landscape is evolving in a very rapid fashion with the emergence of new technologies, new vulnerabilities and new methods of cyberattack tactics. Malicious actors are continually adapting to exploit the latest vulnerabilities and achieve their goals. This constant dynamic makes the traditional cyber security methods based on static rules and signatures increasingly ineffective. The authors underline the importance of combining the existing methods with new forecasting approach.

As a matter of fact, there are valuable proposals to combine and reuse in a novel approach. For instance, the e-Government unified security Framework (e-GUSF) proposed by Hassan and Khalifa in [14] proposes a concrete approach to implement security in government services and is reusable in many contexts. Moreover, Terzi details useful recommendations to protect e-Government from cyber terrorism that tackle multiple dimensions: recruitment of cyber specialists, improve the support of government towards an antivirus software, favor cooperation and establish ethical codes [15].

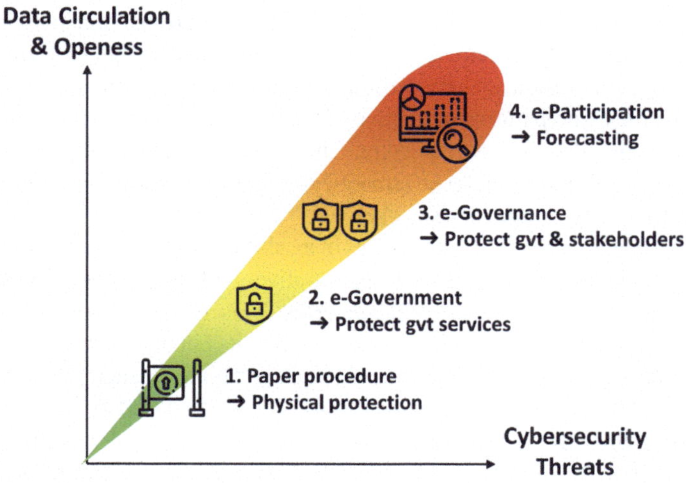

Fig. 6.1 The evolution of cyberthreats compared to the openness of e-Government

Faced with this constantly evolving threat, reactivity is no longer enough. Organizations must now take a proactive approach to security. Dynamic anticipation of cyber-attacks becomes a key element of this approach. By understanding trends and developments in the cyber threat landscape, organizations can anticipate potential attacks and take preventative measures before they occur. Figure 6.1 illustrates the different phases of evolution concerning e-Government.

The reminder of the chapter is organized has follows. Section 6.2 outlines the need for dynamic forecasting in cybersecurity, emphasizing the limitations of traditional reactive approaches. Next, it explores in Sect. 6.3 the potential of epidemiology models drawing parallels between the spread of diseases and the propagation of cyber threats. The analogy between these two phenomena is further elaborated, highlighting their shared characteristics and the potential for cross-disciplinary insights in Sect. 6.4. Thereafter, the authors propose in Sect. 6.5 models that combine epidemiological principles with network analysis and machine learning techniques, offering a comprehensive framework for assessing cyber threats. Finally, we conclude by summarizing the key takeaways and by discussing the promising future of dynamic forecasting models in enhancing cybersecurity.

6.2 Dynamic Forecasting as a Necessity

In the realm of cybersecurity, the concept of dynamic forecasting serves as a pivotal defense strategy, ensuring an anticipatory rather than reactive stance against cyber threats [16]. It hinges upon a continuous, vigilant, and multidimensional analysis of real-time data, leveraging Bayesian probability elements and meticulous monitoring of compromise indicators.

At its core, dynamic forecasting transcends conventional reactive measures, aiming to pre-emptively anticipate and mitigate potential threats that stem from the ever-evolving cyber landscape. By scrutinizing real-time data streams for emerging trends, anomalies, and patterns indicative of suspicious activities, organizations can proactively fortify their security posture. This proactive stance empowers them to adapt swiftly, recalibrating defense strategies in real time and undertaking decisive actions to effectively minimize risks [17].

The efficacy of dynamic forecasting lies in its ability to identify critical vulnerabilities swiftly, enabling organizations to preemptively address potential weak points before they are exploited by cyber adversaries. This methodology functions as a proactive sentinel, not only identifying existing threats but also forecasting potential future risks based on ongoing data analysis and threat intelligence [18].

Moreover, dynamic forecasting serves as a cornerstone in fortifying the resilience of e-Governments and other entities operating in the digital sphere. As these organizations handle sensitive data and critical operations, the proactive adoption of dynamic cyber attack forecasting methods becomes imperative. It's not merely about responding to known threats; rather, it's about preparing for the unknown, the unpredictable nature of cyber threats in an increasingly intricate and perilous digital realm [19].

By integrating dynamic forecasting methodologies into their cybersecurity frameworks, organizations gain a significant advantage in staying ahead of the threat curve. This proactive approach fundamentally alters the cybersecurity paradigm, shifting the focus from reactive defense mechanisms to a proactive, anticipatory strategy that aligns with the rapid evolution of cyber threats [20].

In today's digital landscape, where cyber threats evolve incessantly, the essence of dynamic forecasting in cybersecurity cannot be overstated [21]. It goes beyond traditional reactive measures, instead adopting a proactive stance that anticipates, identifies, and neutralizes potential risks before they materialize into significant threats. By harnessing the power of real-time data analysis and leveraging Bayesian probability elements, dynamic forecasting serves as a preemptive shield against the complexities of cyber threats [22].

This anticipatory approach enables organizations to scrutinize data streams continuously, unveiling emerging trends, anomalies, and suspicious behavioral patterns indicative of potential cyber attacks. It's this proactive scrutiny that empowers entities to fortify their security posture effectively. The agility to adapt and recalibrate defense strategies in real time based on ongoing analysis is pivotal in mitigating risks before they manifest into substantial vulnerabilities [23].

The intrinsic strength of dynamic forecasting lies in its ability to swiftly detect critical vulnerabilities within an organization's cyber infrastructure. By preemptively identifying weak points susceptible to exploitation, organizations can proactively reinforce their defenses, erecting barriers against potential cyber adversaries. Moreover, this methodology isn't merely about addressing current threats; it involves forecasting potential future risks, transforming the reactive approach to a proactive sentinel that guards against unpredictable cyber threats [24].

Proactive anticipation of cyber threats transcends the conventional reactive measures, enabling entities to prepare for the unknown, the uncertainties inherent in the digital sphere's perilous landscape [25].

Integration of dynamic forecasting methodologies into cybersecurity frameworks provides a substantial advantage in combating the ever-evolving threat landscape. This proactive paradigm shift recalibrates the focus from mere reactive defense mechanisms to a forward-looking, anticipatory strategy that resonates with the swift evolution of cyber threats.

In essence, dynamic forecasting signifies a fundamental shift in cybersecurity practices, emphasizing proactive vigilance, adaptability, and the preemptive mitigation of potential risks. Its integration empowers organizations with the agility and foresight required to navigate the intricacies of a constantly evolving digital environment, ensuring robust protection against emerging threats and vulnerabilities.

6.3 Inspirations from Epidemiology

6.3.1 General Considerations

Drawing parallels to the biological realm, there are undeniable similarities between computer viruses and their counterparts in virological transmission. Much like biological viruses that transfer their "genetic" code within a host organism and adapt by mirroring the host's genetic code, there's a phenomenon of hybridization at play [26]. This parallel isn't lost on computer security specialists who are struck by the remarkable analogy. In the context of cyberattacks, when a virus seeks to infect a system, it often follows a specific *modus operandi*. Initially, it manifests itself as a concise set of binary machine language code lines, which are then affixed onto a program used by the target system. This code seeks to modify the program's behavior in a way that aligns with the virus's objectives. The virus might either be entirely self-contained within the plugin or serve as a primer, responsible for downloading a more extensive program that assumes the role of the actual virus. Once implanted into a host program, the plugin typically propagates itself to other programs within the system, intensifying the contamination, and making disinfection an increasingly formidable challenge [27].

The nomenclature likening this type of program to a virus is far from arbitrary, as the similarities are profound. Just as biological viruses operate in strikingly analogous manners, with a genetic material encapsulated in an envelope, which allows them to withstand the external environment before infiltrating the host organism, computer viruses also share these characteristics. When a virus penetrates a host cell, it introduces a fragment of DNA or RNA, akin to a "genetic program", that commandeers the host's cellular machinery to infiltrate other cells.

In the words of David Evans, as elaborated in "What biology can (and can't) teach us about security", a computer virus faces a significant challenge akin to its biological counterpart—evading the host's immune system. To survive and propagate, viruses, both biological and digital, employ strategies grounded in mutations and polymorphism [28]. Put simply, viruses adapt to their environment to conceal their presence more effectively. Just as biological viruses employ incubation periods as a survival strategy, computer viruses often exhibit a period during which they remain relatively innocuous to the host organism, making it easier for healthy carriers to unwittingly spread the infection [29]. An effective virus, whether biological or digital, avoids being readily detectable and disseminates within the host system without causing conspicuous harm [30]. This intricate dance between viruses and their hosts, whether biological or digital, underscores the critical need for dynamic cyberattack forecasting methods in the increasingly complex and perilous digital realm, especially for e-Governments looking to safeguard their data, operations, and reputation in this ever-evolving landscape.

There is a second striking similarity which is that of the very definition of what a virus is: actually, deciding what program behaviors are malicious is highly subjective. For example, a program that replaces advertisement in web pages maybe a malware, maybe an ad-blocker behavior than the machine owner desires, or maybe a required trade-off the user knowingly accepts [31].

The question of whether or not an organism can be considered a virus is also a question for biologists that is not as obvious as it seems. Indeed, we cannot simply limit the definition of a virus to its malicious nature, a certain number of viruses actually have a positive impact on the human body. What determines whether a virus is good or bad for the host organism is simply the damage caused. For example, when an organization's computer network is protected by a sufficiently up-to-date anti-virus system, it may be a good idea for computers to come into regular contact with viruses to test its effectiveness in detecting and quarantining this type of virus. Anti-virus software does not have the same adaptation capabilities as the human body's immune system: the anti-virus will not be able to detect a virus unless its signature appears in its signature database [32].

6.3.2 Illustrating Example

To gain a better understanding of the analogy between computer viruses and biological viruses, let's begin with an example. In its December 2023 issue, Wired Magazine unveiled the story of the MIRAI malware[1] that was also depicted in [33]. This malicious software targeted small devices such as routers and other IoT devices. The impact of this malware and its variants was significant, causing disruptions like blocking access to major websites such as the New York Times and Twitter, and affecting companies like Amazon and Netflix. Initially infecting just a few devices, it rapidly spread to 650,000 devices at an exponential rate with major costs, similarly to previous cyber attacks [34]. The spread of infected devices resembled the pace of

[1] https://www.wired.com/story/mirai-untold-story-three-young-hackers-web-killing-monster/.

highly aggressive diseases like Covid-19 (a direct virus) or malaria (a parasite with a vector).

Not all devices were susceptible to infection; there existed a subset that could be infected. However, being susceptible was not sufficient for infection; the device also needed to be accessible for a period. Sometimes, due to poor broadband connections and frequent disconnections, a susceptible device might not become contaminated. Therefore, we can identify at least three phases in this process: susceptibility, exposure, and infection.

Similar to traditional epidemiology, defending organizations established honeypots that simulated devices to monitor malware. This approach allowed them to create a map of the malware's spread, providing valuable data to understand the epidemic's characteristics. This served a dual purpose: first, in finding a cure (in this context, a method for recovery), and second, in creating a patch to block the malware, essentially a vaccine.

Additionally, as we will explore later, the spread of malware follows steps that have been described and analyzed in epidemiology. However, there are notable differences, particularly in the mechanism of infection. Unlike natural diseases, there are no mosquitoes or water droplets involved. In the case of MIRAI, as reported by Wired, the hackers devised a centralized three-step structure. Their infected devices would scan for other susceptible devices and then report the vulnerable ones they discovered to a server. This server would exploit the devices through telnet to install the malware. Subsequently, a separate command-and-control server would oversee these malware-infected bots, periodically issuing new commands to determine which targets to attack. This demonstrates that there is intent and techniques employed by attackers that do not exist in nature.

6.4 Analogy Between the Spread of Diseases and the Spread of Cyber Attacks

6.4.1 Pathogen Versus Malware

In the realm of biological epidemiology, pathogens serve as the primary organisms responsible for causing diseases within living organisms. Similarly, within the domain of cybersecurity, the role of pathogens is metaphorically assumed by malware—malicious software entities meticulously designed to infiltrate and compromise computer systems. Much like how pathogens invade and affect the biological structure of a host, malware, including viruses or Trojan horses, infiltrates computer systems with the intent to disrupt, manipulate, or extract sensitive data. The parallel lies in their ability to infiltrate and cause disruptions within their respective environments, albeit in different contexts—one biological and the other digital.

This comparison highlights the resemblances between the mechanisms of transmission and impact of pathogens in biological settings and the actions of malware within the intricate landscape of cybersecurity. Just as pathogens circulate within a biological host, causing harm and disruptions to physiological systems, malware circulates within computer networks, exploiting vulnerabilities and causing disruptions to digital functionalities. Both entities, pathogens and malware, possess the ability to spread and propagate within their respective ecosystems, causing varying degrees of damage and disruption [35].

6.4.2 Susceptible Host Versus Vulnerable System

In the context of biological epidemiology, a susceptible host refers to an individual who lacks immunity or defenses against a particular disease, making them prone to infection or illness. Similarly, within cybersecurity frameworks, a vulnerable computer system embodies the concept of susceptibility to malware infection. Just as a susceptible host is at risk of contracting a disease due to a compromised immune system or exposure to infectious agents, a vulnerable computer system is at risk of malware infiltration due to existing security weaknesses or unpatched vulnerabilities.

Vulnerabilities within computer systems, akin to compromised immune systems in susceptible hosts, serve as entry points for cyber attackers seeking to exploit these weaknesses. These vulnerabilities can manifest as unpatched security flaws, misconfigurations, or outdated software, creating loopholes through which malware can infiltrate and compromise the system's integrity. Just as a compromised immune system makes an individual susceptible to a range of diseases, these system vulnerabilities render them susceptible to cyber attacks [36].

The analogy emphasizes the criticality of addressing vulnerabilities within computer systems to mitigate the risk of malware infiltration and subsequent compromise of sensitive data or system functionalities. By remedying these vulnerabilities through robust security measures, including timely software updates, security patches, and proactive threat monitoring, organizations can significantly reduce the susceptibility of their systems to cyber attacks, thereby fortifying their digital defenses and enhancing cybersecurity resilience [37].

6.4.3 Spread and Contagion

In the context of biological epidemiology, the spread of diseases occurs through various mediums, including direct contact between individuals, airborne transmission, or vectors like insects or animals. Similarly, within the realm of cybersecurity, cyber attacks propagate from one system to another, exploiting various infection vectors, including phishing emails, interconnected computer networks, or compromised devices.

Much like how diseases disseminate from one host to another through contact or vectors, cyber attacks traverse through digital networks, aiming to infiltrate interconnected systems or devices. Cyber attackers capitalize on vulnerabilities within these networks or leverage deceptive tactics like phishing emails to gain access to unsuspecting systems. Just as diseases exploit opportunities to spread through physical interactions or carriers, cyber attackers capitalize on digital pathways and vulnerabilities to proliferate their attacks across interconnected systems or networks.

The comparison underscores the parallels between the transmission dynamics of diseases and the propagation methods of cyber attacks. Both phenomena rely on pathways or vectors to propagate, whether it's physical contact, intermediaries, or digital pathways. Understanding these parallels aids in comprehending the mechanisms through which cyber attacks traverse digital landscapes, highlighting the significance of fortifying defenses against potential infection vectors to prevent the spread and proliferation of cyber threats [38].

6.4.4 Strain Evolution

In the realm of biological epidemiology, pathogens undergo evolutionary changes over time to evade the host's immune responses. Similarly, within the sphere of cybersecurity, cyber attacks exhibit a similar evolutionary pattern, wherein malware variants continuously evolve, becoming more sophisticated and challenging to detect by conventional security measures.

Much like how biological pathogens mutate to circumvent immune defenses, cyber attackers constantly refine and enhance malware variants to bypass security protocols and evade detection mechanisms. These advancements in malware design and sophistication aim to exploit new vulnerabilities, outmaneuver existing defense systems, and adapt to the evolving cybersecurity landscape [39].

The evolution of malware variants reflects a strategic response by cyber attackers to counteract the advancements in cybersecurity measures. Just as pathogens evolve to overcome the host's immune responses, cyber attackers evolve malware to circumvent and exploit the defenses implemented by organizations and security systems [40].

This comparison underscores the dynamic nature of both biological pathogens and cyber threats, emphasizing the continual evolution and adaptation exhibited by malicious entities. Understanding this evolution is critical in enhancing cybersecurity strategies to anticipate and counteract the evolving tactics employed by cyber attackers, thereby fortifying defenses against increasingly sophisticated cyber threats and minimize the risks associated with ever-evolving cyber attacks [41–43].

6.5 Proposition of Integrated Models to Assess Cyber-Threats

The realm of cybersecurity is constantly evolving, marked by an intricate dance between cyber threats and defense mechanisms. In this section, we present a pioneering effort to integrate insights from epidemiology, a field traditionally associated with the study of disease dynamics, into the assessment of cyber-threats. Our goal is to offer a holistic perspective that captures the contagious nature of cyber threats and the evolving landscape of protection. This approach is the opposite slope of multiple proposals coming from ICT to struggle against the Covid-19 pandemic [44, 45].

6.5.1 The SEIR Epidemiological Model for Cybersecurity

In the pursuit of a robust model to assess cyber-threats, we turn to the Susceptible-Exposed-Infectious-Recovered (SEIR) epidemiological framework as depicted by Hethcote in [46]. Adapted for the cybersecurity domain, this model incorporates the concept of latency, where systems may be exposed to threats before becoming actively compromised. By drawing parallels between the spread of diseases and the propagation of cyber threats, the SEIR model provides a nuanced lens through which to understand the dynamics of susceptibility, exposure, compromise, and recovery in complex cyber ecosystems.

This integrated model seeks to contribute significantly to the understanding of cyber threats by bridging the gap between epidemiological concepts and cybersecurity dynamics. As the digital landscape continues to evolve, a multidimensional perspective becomes crucial for effective threat assessment and mitigation. Through this integrated model, we aim to provide a tool that not only dissects the intricacies of cyber threats but also offers a foundation for proactive cybersecurity strategies informed by principles rooted in epidemiological insights.

The SEIR (Susceptible-Exposed-Infectious-Recovered) epidemiological model, introduces an "exposed" compartment representing systems that are in the latency period after being exposed to a threat before becoming infectious (Fig. 6.2).

Let's consider the following populations:

- S(t): the population of susceptible systems at time t
- E(t): the population of exposed systems at time t
- I(t): the population of infectious systems at time t
- R(t): the population of recovered/isolated systems at time t

Mathematical Formulation

We propose the following equations inspired by the SEIR epidemiological model:

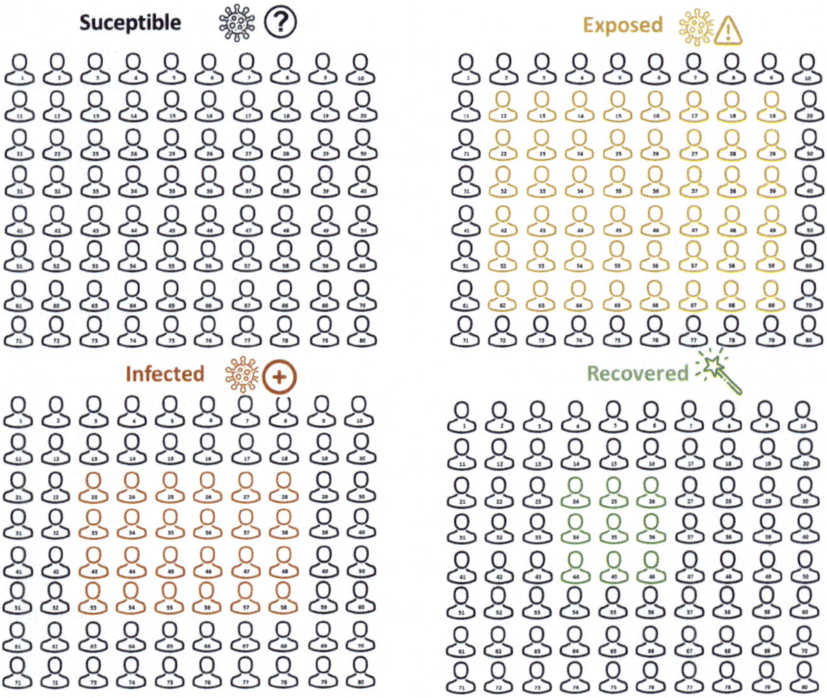

In this simple example, the parameters of the SEIR model have the following characteristics:

- β is high which means that the cyberattack has a high rate of propagation
- σ is high which means that exposed systems are rapidly infected
- γ is moderate which means that the recovery is not immediate

Fig. 6.2 Illustration of the SEIR model applied to cybersecurity

$$\frac{dS}{dt} = -\beta \cdot S(t) \cdot I(t)$$

$$\frac{dE}{dt} = -\beta \cdot S(t) \cdot I(t) - \sigma \cdot E(t)$$

$$\frac{dI}{dt} = \sigma \cdot E(t) - \gamma \cdot I(t)$$

$$\frac{dR}{dt} = \gamma \cdot I(t)$$

Interpretation

- The first equation represents the rate at which susceptible systems (S) become infected (I).

- The second equation represents the rate at which exposed systems (E) increase due to new infections and decrease as they transition to the infectious state.
- The third equation represents the rate at which infectious systems (I) increase due to the transition from the exposed state and decrease due to recoveries.
- The fourth equation represents the rate at which recovered/isolated systems (R) increase due to recoveries.

Adaptation to Cybersecurity

- S can represent the population of systems susceptible to be attacked.
- E can represent the population of exposed systems in the latency period after being exposed to a threat.
- I can represent the population of infected systems.
- R can represent the population of recovered or isolated systems.

Parameters

- β (transmission rate),
- σ (latent period),
- γ (recovery/isolation rate)

The parameters should be adapted based on the characteristics of the unique cybersecurity environment.

Algorithmic Implementation

It is then, easy to implement the model algorithmically using numerical methods, such as the Euler method, to simulate the propagation of cybersecurity threats over time. This method was described by Dupuis and Nagurney in [48] and applied in a similar context in [49] by Nagurney and Shukla. Originally, this model introduces a latent period, allowing for the consideration of systems that are exposed to a threat but not immediately infectious.

The applicability of the proposal is detailed in Table 6.1.

Table 6.1 Evaluation of the first proposal

Maturity of the proposed method	High. The method is largely implemented in the literature. This facilitates its immediate applicability
Time frame of expected availability of proposed method	Almost Immediate. It only requires to adjust the parameters of the method
Requirements to apply proposed method	Very limited. As the approach is based on statistics it does not necessarily require to obtain complex data from the different systems
Expected efficiency of the proposed method	Intermediary. The method helps to predict the effects of an attack and to define counter-measures but it is not specific to different attackers and/or targets

6.5.2 Cybersecurity Propagation Model: Inspired by Herd Immunity and Cyber Immunity

Herd immunity occurs when a significant portion of a population becomes immune to a disease, reducing its spread. It is achieved through vaccination or natural infection, providing collective protection and safeguarding vulnerable individuals. In cybersecurity, the concept of "cyber immunity" draws parallels, suggesting that the presence of protected systems can confer resilience to the entire network, disrupting the propagation of cyber threats and providing a layer of defense against potential attacks. Applying herd immunity principles to cybersecurity strategies emphasizes the importance of comprehensive protection measures to create a resilient and interconnected defense against evolving cyber threats.

Building upon the foundations laid by epidemiological concepts, we introduce a novel Cybersecurity Propagation Model, inspired from the SIR model described in [47] whose popularity was renewed in the Covid-19 era. This model is inspired by the notions of herd immunity, where the presence of protected entities confers resilience to the entire system, and cyber immunity, a dynamic factor that evolves in response to cyber contagion. By infusing these concepts into the model, we seek to capture the interplay between susceptible, compromised, and immune systems, offering a fresh perspective on the dynamics of cyber-threat propagation.

For our cybersecurity propagation model, let's consider the following populations:

- S(t): the population of susceptible systems at time t.
- I(t): the population of infected (compromised) systems at time t.
- R(t): the population of recovered systems with cyber immunity at time t.

Mathematical Formulation

The model equations are inspired by epidemiological principles:

$$\frac{dS}{dt} = -\beta \cdot S(t) \cdot I(t)$$

$$\frac{dI}{dt} = \beta \cdot S(t) \cdot I(t) - \alpha \cdot I(t) \cdot R(t)$$

$$\frac{dR}{dt} = \alpha \cdot I(t) \cdot R(t)$$

Interpretation

- The first equation represents the rate at which susceptible systems (S) become compromised (I).
- The second equation represents the rate at which infected systems (I) increase due to new compromises and decrease due to the presence of cyber immunity due to the recovery (R).

- The third equation represents the rate at which systems acquire cyber immunity (R) due to exposure to compromised systems and further recovery.

Adaptation to Cybersecurity

- S can represent the population of systems without adequate security measures.
- I can represent the population of compromised systems.
- R can represent the population of systems with developed cyber immunity.

Parameters

We can adjust the parameters (β, α) based on the characteristics of the specific cybersecurity environment, where:

- β: Compromise rate due to interactions with compromised systems.
- α: Rate of development of cyber immunity.

Algorithmic Implementation

It is then easy to implement the model algorithmically using numerical methods to simulate the dynamic interactions between susceptible, compromised, and immune systems over time.

This model directly incorporates the concept of cyber immunity, analogous to herd immunity in epidemiology. The presence of cyber immunity (R) reduces the spread of compromises within the system.

Considerations

Cyber immunity could be influenced by factors like the effectiveness of security measures, user awareness, and incident response capabilities. Adapt the model to account for the specific characteristics and nuances of your cybersecurity landscape.

This model aligns more closely with epidemiological concepts by introducing the idea of immunity and contagion, making it a more nuanced representation of the dynamics of cybersecurity threats.

The applicability of the proposal is detailed in Table 6.2.

6.5.3 Multi-level SEIR Epidemiological Model for Cybersecurity

In this approach, we propose to adjust the SEIR model to consider two additional dimensions (Fig. 6.3):

1. the difference of maturity and protection according to the considered stakeholder (government service, private operator, citizen…)
2. the difference in the threats according to their probability and their potential impact

Table 6.2 Evaluation of the second proposal

Maturity of the proposed method	High. The method is largely implemented in the literature. This facilitates its immediate applicability
Time frame of expected availability of proposed method	Immediate. It only requires to adjust the parameters of the method. The parameters are reduced compared to the first method
Requirements to apply proposed method	Very limited. As the approach is based on statistics it does not necessarily require to obtain complex data from the different systems
Expected efficiency of the proposed method	Low. The method helps to predict the effects of an attack and to define counter-measures but it is not specific to different attackers and/or targets. It is less efficient than the first method

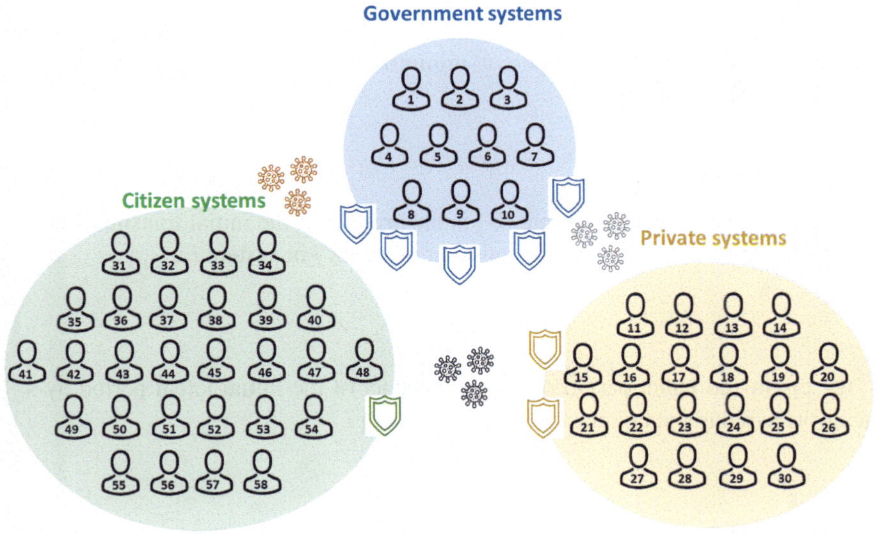

Fig. 6.3 Illustration of a Multi-SEIR model applied to cybersecurity

Parameters

- i (system) and j (cyber-attack)
- $\beta_{i,j}$ (transmission rate of cyber-attack *j* on system *i*),
- $\sigma_{i,j}$ (latent period in system *i* after cyber-attack *j*),
- $\gamma_{i,j}$ (recovery/isolation rate of system *i* after cyber-attack *j*).

To simplify the modelization, we can define that the systems are equivalent for each stakeholder.

Mathematical Formulation

$$\frac{dS}{dt} = -\sum_{i,j} \beta_{i,j} \cdot S(t) \cdot I(t)$$

$$\frac{dI}{dt} = \sum_{i,j} \beta_{i,j} \cdot S(t) \cdot I(t) - \alpha_{i,j} \cdot I(t) \cdot R(t)$$

$$\frac{dR}{dt} = \sum_{i,j} \alpha_{i,j} \cdot I(t) \cdot R(t)$$

Considerations

Cyber security requires to increase the level of protection of each link in the chain. The weakest one is the most dangerous. This motivates the public policies towards enhancing the skills of all the ICT ecosystem whether it is governmental, private or personal.

A key question is how to optimize the allocation of means to enhance the cyber protection. For example, we propose to minimize the infection of systems according to a limited effort. Define:

$$\beta'_{i,j} = \beta_{i,j} \times \mathrm{b}'_{i,j}$$

the optimized transmission rate $\beta'_{i,j}$ is the product of the initial transmission rate by $\mathrm{b}'_{i,j}$ the effort to reduce the transmission of attack *j* in system *i*.

$$\sigma'_{i,j} = \sigma_{i,j} \times \mathrm{s}'_{i,j}$$

the optimized latent period $\sigma'_{i,j}$ is the product of the initial latent period by $\mathrm{s}'_{i,j}$ the effort to delay the effects of attack *j* in system *i*.

$$\gamma'_{i,j} = \gamma_{i,j} \times \mathrm{g}'_{i,j}$$

the optimized recovery rate $\gamma'_{i,j}$ is the product of the recovery rate by $\mathrm{g}'_{i,j}$ the effort to hasten the recovery of attack *j* in system *i*.

We propose to solve the following problem:

$$arg \min_{\substack{\{\mathrm{b}'_{i,j}\} \\ \{\mathrm{s}'_{i,j}\} \\ \{\mathrm{g}'_{i,j}\}}} \sum_{t=0..N} I(t)$$

Under the constraint that the effort of a system *i* is limited by a maximum value Vi

Table 6.3 Evaluation of the third proposal

Maturity of the proposed method	High. The convex optimization and the SEIR methods are well-known and largely implemented in the literature. However, the combination of both methods requires high expertise
Time frame of expected availability of proposed method	Intermediate. It requires to adjust multiple parameters as well as convex optimization. There is a real effort of implementation
Requirements to apply proposed method	Moderate. As the approaches are well-known, the requirements may be easily fulfilled
Expected efficiency of the proposed method	Very high. The method is very useful to optimize the protection and forecast multiple and complex cyber attacks

$$\forall i \sum_j \mathrm{b}'_{i,j} + \mathrm{s}'_{i,j} + \mathrm{g}'_{i,j} \leq V_i$$

Thus, the key objective is to experiment and adjust a strategy for each system to invest in the dimension that reduces the most the total infection.

Algorithmic Implementation

The problem exposed may be easily solved by using convex optimization as described in [50] and illustrated in [51].

Perspectives

Another interesting approach consists in grouping multiple systems so that the question is how to optimize them simultaneously. A key question of CTO is to identify whether it is a better idea to have a transversal approach or to protect only critical systems. The applicability of the proposal is detailed in Table 6.3.

6.5.4 Potential Integration for Comprehensive Threat Assessment

The integration of the SEIR model and the Cybersecurity Propagation Model aims to provide a comprehensive framework for assessing cyber-threats.

By proposing the temporal aspects of exposure and recovery from the SEIR model and the immunity dynamics from the Cybersecurity Propagation Model, we strive to capture a more realistic representation of the cyber-threat landscape. The needed integrated approach acknowledges the complexity of cyber ecosystems, where threats are not isolated events but dynamic processes that unfold over time (Fig. 6.4).

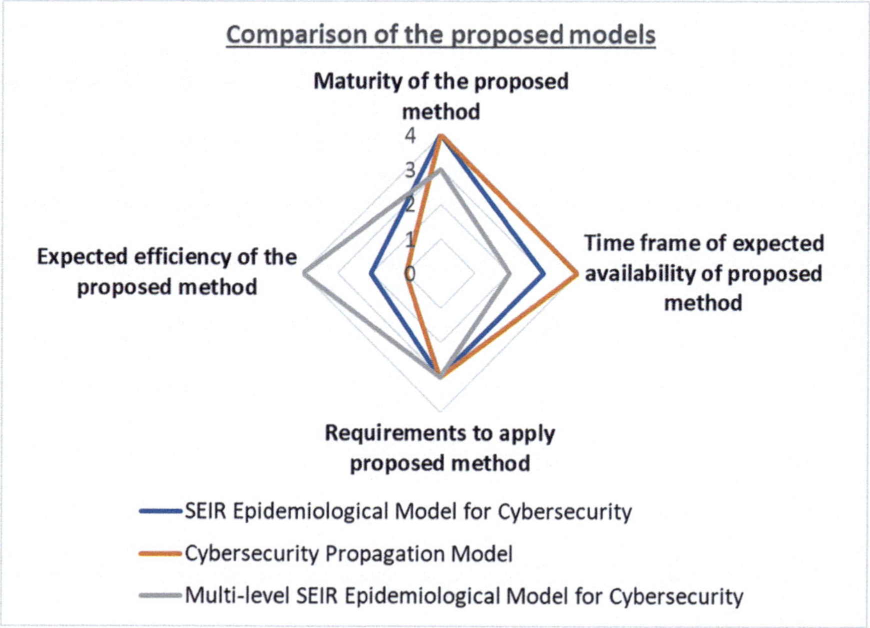

Fig. 6.4 Comparison of the different models inspired from epidemiology to enhance the cybersecurity of e-Government

6.6 Conclusion and Perspectives

The integration of epidemiology-inspired models into cybersecurity strategies for eGovernment marks a paradigm shift in addressing the evolving landscape of digital threats.

The SEIR model, traditionally used in epidemiology, represents Susceptible, Exposed, Infected, and Recovered states. Applied to cybersecurity, this model maps susceptibility to cyber threats, exposure to vulnerabilities, active infection of systems, and the recovery or mitigation processes. By adapting the SEIR model to the digital realm, governments gain a powerful tool for forecasting and mitigating cyber threats. This model allows for the identification of vulnerable states within the e-Government infrastructure, enabling preemptive measures to bolster cybersecurity protocols. The SEIR model offers a structured approach to incident response in e-Government. By categorizing cyber threats into distinct stages, authorities can allocate resources effectively, prioritize responses, and streamline recovery efforts.

Drawing inspiration from epidemiological concepts of herd immunity, the Cybersecurity Propagation Model introduces the idea that a sufficiently immune digital environment can protect against widespread cyber threats. As a certain percentage of systems achieve cyber immunity, the overall risk to the e-Government ecosystem diminishes.

Governments can use the Cybersecurity Propagation Model to devise strategies for enhancing cyber immunity. This involves not only fortifying individual systems but also fostering collaborative efforts between government entities, creating a network effect that elevates overall cybersecurity resilience.

The Cybersecurity Propagation Model emphasizes the importance of collective defense. e-Government agencies can collaborate to share threat intelligence, implement standardized security measures, and collectively raise the immunity threshold, thereby reducing the likelihood and impact of cyber threats.

The adoption of epidemiology-inspired models, such as the SEIR model and the Cybersecurity Propagation Model, underscores the proactive approach governments can take in securing digital governance. These models provide a structured framework for understanding, predicting, and mitigating cyber threats in the context of e-Government. By embracing these innovative tools, governments can fortify their digital infrastructure, enhance collaboration, and ensure the continued progress of governance in the digital age.

This paper also describes innovative approaches such as the Multi-level SEIR Epidemiological Model for Cybersecurity. As a matter of fact, the authors not only propose to adapt existing models to cybersecurity of e-Government but also to imagine new models based on the specific context. A key concern of civil servant in the IT domain is not to define a strategy but rather to obtain resources to achieve the best possible protection of public services. As a consequence, future research will be dedicated to define a method for providing the best security measures at the lowest cost. Some concrete experiments will also be conducted to verify the effectivity of the proposed approaches.

References

1. Barthelmess, B., & Langlois, J. (2018). *Digital identity: Legal and economic issues.*
2. Grönlund, Å., & Horan, T. (2005). Introducing e-Gov: History, definitions, and issues. *Communications of the Association for Information Systems, 15.* https://doi.org/10.17705/1CAIS.01539
3. Shareef, S. (2016). Enhancing security of information in e-Government. *Journal of Emerging Trends in Computer Information Science, 7*(3), 139–146.
4. Santamaría-Philco, A., Canós Cerdá, J. H., & Penadés Gramaje, M. C. (2019). Advances in e-participation: A perspective of last years. *IEEE Access, 7,* 155894–155916. https://doi.org/10.1109/ACCESS.2019.2948810.
5. Singh, S., Karaulia, D.S. (2011). E-governance: Information security issues. In *Proceedings of the international conference on computer science and information technology* (pp. 120–124).
6. Nikolopoulos, K., Patrikakis, C. Z., & Lin, B.-S. (2004). Forecasting systems for e-Government. *Electronic Government, 1*(4), 374–383.
7. Reitz, J. C. (2006). e-Government. *American Journal of Comparative Law, 54.* U Iowa Legal Studies Research Paper No. 05–43. https://ssrn.com/abstract=887664
8. Alshehri, M. A., & Drew, S. (2010). Implementation of e-Government: Advantages and challenges. *International Conference E-Activity and Leading Technologies 2010,* 79–86. http://hdl.handle.net/10072/40620

9. Buccafurri, F., Fotia, L., & Lax, G. (2015). A privacy-preserving e-participation framework allowing citizen opinion analysis. *Electronic Government, an International Journal, 11*(3), 185–206.
10. Efe, A., & Kazdal, H. (2019). IT security trends for e-Government threats. *International Journal of Multidisciplinary Studies and Innovative Technologies, 3*(2), 105–110. https://doi.org/10.1063/5.0128315
11. Mishra, S., et al. (2021). Impact of security standards and policies on the credibility of e-Government. *Journal of Ambient Intelligence and Humanized Computing*. https://doi.org/10.1007/S12652-020-02767-5
12. Gerunov, A. (2020). Attitudes towards privacy by design in e-Government: Views from the trenches. *Journal Of Social And Administrative Sciences, 7*(1), 1–17. https://doi.org/10.1453/jsas.v7i1.2028
13. FBI Internet Crime Report 2022 (PDF). (2023). Internet Crime Complaint Center. https://www.ic3.gov/Media/PDF/AnnualReport/2022_IC3Report.pdf
14. Hassan, R. G., & Khalifa, O. O. (2016). e-Government—An information security perspective. *International Journal of Computing Trends Technology., 36*(1), 1–9.
15. Terzi, M. (2019). e-Government and cyber terrorism: Conceptual framework, theoretical discussions and possible solutions. *Turkish Journal of TESAM Academy, 6*(1), 213–247. https://doi.org/10.30626/tesamakademi.528011
16. Camp, L. J., Grobler, M., Jang-Jaccard, J., Probst, C., Renaud, K., & Watters, P. (2019). Measuring human resilience in the face of the global epidemiology of cyber attacks. In: Proceedings of the 52nd Hawaii International Conference on System Sciences, MAUI United States, 8 January 2019 (pp. 4763–4772). https://doi.org/10.24251/HICSS.2019.574
17. Cohen, F. (1987). Computer viruses: Theory and experiments. *Computers & Security, 6*, 22–35. https://doi.org/10.1016/0167-4048(87)90122-2
18. Moradzadeh, A., Mohammadpourfard, M., Genc, I., Şeker, ŞS., & Mohammadi-Ivatloo, B. (2022). Deep learning-based cyber resilient dynamic line rating forecasting. *International Journal of Electrical Power & Energy Systems, 142*, 108257.
19. Alazab, A., Abawajy, J., Hobbs, M., Layton, R., & Khraisat, A. (2013) Crime toolkits: The productisation of cybercrime. In *2013 12th IEEE international conference on trust, security and privacy in computing and communications, Melbourne*, 16–18 July 2013 (pp. 1626–1632). https://doi.org/10.1109/TrustCom.2013.273
20. Gross, G. (2020). Intrusion Detection Techniques Methods and Best Practices. AT&T Business. https://cybersecurity.att.com/blogs/security-essentials/intrusion-detection-techniques-methods-best-practices
21. Jean, L. (2021). When linguistics meets computer science: Stylometry and professional discourse. *Training, Language and Culture, 5*(2), 51–61. https://doi.org/10.22363/2521-442X-2021-5-2-51-61
22. Antonakakis, M., Perdisci, R., Lee, W., Vasiloglou, N., & Dagon, D. (2011). Detecting malware domains at the upper DNS hierarchy. *USENIX Security Symposium, 11*, 1–16.
23. Biggio, B., & Roli, F. (2018). Wild patterns: Ten years after the rise of adversarial machine learning. *Pattern Recognition, 84*, 317–331. https://doi.org/10.1016/j.patcog.2018.07.023
24. Gil, S., Kott, A., & Barabási, A. L. (2014). A genetic epidemiology approach to cyber-security. *Scientific Reports, 4*, Article No. 5659. https://doi.org/10.1038/srep05659
25. Goyal, S., Jabbari, S., Kearns, M., Khanna, S., & Morgenstern, J. (2016). Strategic network formation with attack and immunization. *International Conference on Web and Internet Economics, Montréal, 11–14*, 429–443. https://doi.org/10.1007/978-3-662-54110-4_30
26. Novick, L. F. (2005). Epidemiologic approaches to disasters: Reducing our vulnerability. *American Journal of Epidemiology, 162*, 1–2. https://doi.org/10.1093/aje/kwi164
27. Feng, C., Qin, Z., Cuthbet, L., & Tokarchuk, L. (2008). Propagation model of active worms in P2P networks. In *The 9th International Conference for Young Computer Scientists, Hunan, 18–21 November 2008* (pp. 1908–1912). https://doi.org/10.1109/ICYCS.2008.237
28. Evans, D. (2004). What biology can (and can't) teach us about security. In *13th USENIX Security Symposium (USENIX Security 04)*.

29. Koroniotis, N., Moustafa, N., & Sitnikova, E. (2020). A new network forensic framework based on deep learning for internet of things networks: A particle deep framework. *Future Generation Computer Systems, 110*, 91–106. https://doi.org/10.1016/j.future.2020.03.042
30. Ivenson, S. (2020). *IP time to live (TTL) and hop limit basics-packet pushers.*
31. Jerkins, J. A., & Stupiansky, J. (2018). Mitigating IoT insecurity with inoculation epidemics. In *Proceedings of the ACMSE 2018 conference, association for computing machinery, New York, 29 March 2018* (pp. 1–6). https://doi.org/10.1145/3190645.3190678
32. Brantly, A. (2017). Public health and epidemiological approaches to national cybersecurity: A baseline comparison. In *US national cybersecurity: International politics, concepts and organization, Routledge, London, Chapter 7*. https://doi.org/10.4324/9781315225623-7
33. Sinanović, H., Mrdovic, S. (2017). Analysis of Mirai malicious software. In *2017 25th international conference on software, telecommunications and computer networks (SoftCOM), Split, Croatia, 2017* (pp. 1–5). https://doi.org/10.23919/SOFTCOM.2017.8115504
34. Anderson, R., Barton, C., Rainer, B., Clayton, R., Ga, C., Grasso, T., Levi, M., Moore, T., & Vasek, M. (2019). Measuring the changing cost of cybercrime our framework for analysing the costs of cybercrime. In *Workshop on the economics of information security (WEIS)* (pp. 1–32).
35. Modini, J., Lynar, T., Sitnikova, E., & Joiner, K. (2020). Applications of epidemiology to cybersecurity. In *European conference on cyber warfare and security* (pp. 483–490). Academic Conferences International Limited.
36. Liu, Z., Zeng, Y., Zhang, P., Xue, J., Zhang, J., & Liu, J. (2018). An imbalanced malicious domains detection method based on passive DNS traffic analysis. *Security and Communication Networks*, Article ID: 6510381. https://doi.org/10.1155/2018/6510381
37. Marsden, T., Moustafa, N., Sitnikova, E., & Creech, G. (2017). Probability risk identification based intrusion detection system for SCADA systems. In *International conference on mobile networks and management* (pp. 353–363). Springer. https://doi.org/10.1007/978-3-319-90775-8_28
38. Rothman, K. J. (2012). *Epidemiology: An introduction*. Oxford University Press.
39. LaMorte, W. (2019). *The evolution of epidemiologic thinking*. Boston University School of Public Health.
40. Langlois, J. (2021). *Evaluating and insuring cyber risks within organizations (Doctoral dissertation, EHESS).*
41. Sun, C. A., Hahn, A., & Liu, C. C. (2018). Cyber security of a power grid: State-of-the-art. *International Journal of Electrical Power and Energy Systems, 99*, 45–56. https://doi.org/10.1016/j.ijepes.2017.12.020
42. Veksler, V. D., Buchler, N., Hoffman, B. E., Cassenti, D. N., Sample, C., & Sugrim, S. (2018). Simulations in cyber-security: A review of cognitive modeling of network attackers, defenders, and users. *Frontiers in Psychology, 9*, 691. https://doi.org/10.3389/fpsyg.2018.006
43. Singh, M., Singh, M., & Kaur, S. (2019). Detecting bot-infected machines using DNS fingerprinting. *Digital Investigation, 28*, 14–33. https://doi.org/10.1016/j.diin.2018.12.005
44. Gaie, C., & Mueck, M. (2022). An artificial intelligence framework to ensure a trade-off between sanitary and economic perspectives during the COVID-19 pandemic. In *Deep learning for medical applications with unique data* (pp. 197–2017). Academic Press, ISBN 9780128241455. https://doi.org/10.1016/B978-0-12-824145-5.00008-3
45. Gaie, C., & Mueck, M. (2022). COVID-19 lethality reduction using artificial intelligence solutions derived from telecommunications systems. In *Data science for COVID-19* (pp. 647–666). Academic Press, ISBN 9780323907699. https://doi.org/10.1016/B978-0-323-90769-9.00025-6
46. Hethcote, H. W. (2000). The mathematics of infectious diseases. *SIAM Review, 42*, 599–653. https://doi.org/10.1137/S0036144500371907
47. Hethcote, H. W. (1989). Three basic epidemiological models. In S.A. Levin, T.G. Hallam, & L.J. Gross (Eds.), *Applied mathematical ecology*. Biomathematics (vol. 18). Springer. https://doi.org/10.1007/978-3-642-61317-3_5
48. Dupuis, P., & Nagurney, A. (1993). Dynamical systems and variational inequalities. *Annals of Operations Research, 44*, 9–42. https://doi.org/10.1007/BF02073589

49. Nagurney, A., & Shukla, S. (2017). Multifirm models of cybersecurity investment competition vs. cooperation and network vulnerability. *European Journal of Operational Research, 260*(2), 588–600. https://doi.org/10.1016/j.ejor.2016.12.034
50. Boyd, S., & Vandenberghe, L. (2004). *Convex optimization.* Cambridge University Press.
51. Gaie, C., Assaad, M., Muck, M., & Duhamel, P. (2008). Distributed discrete resource optimization in heterogeneous networks. In *2008 IEEE 9th workshop on signal processing advances in wireless communications, Recife, Brazil* (pp. 560–564). https://doi.org/10.1109/SPAWC.2008.4641670

Chapter 7
The Provision of e-Services by Public Administration Bodies and Their Cybersecurity

Christophe Gaie and Mirosław Karpiuk

Abstract Electronic services in information society provide significant opportunities of meeting the needs of such society, including those fulfilled by public administration. However, in order for e-administration to provide public services in an efficient manner, it needs to keep up with technological developments, thus meeting the technical standards envisaged for such services. Services in the digital sector can change greatly and rapidly, so public administration needs to monitor and analyse them on an ongoing basis and, where necessary, adapt to such changes. There are certain legal obstacles that hamper the development of public e-services and the efficient functioning of e-administration, especially in the competitive market of information society services. Administration bodies are obliged to act in compliance with the law and within its limits, also in the electronic market which requires considerable mobility. At the same time, they must keep up with the dynamically changing needs of the participants in that market and the freedom of decision that applies there, whereas public administration itself does not enjoy such freedom and can only take measures that are prescribed by law.

Keywords e-Administration · Public e-services · Cybersecurity

7.1 Definition of e-Administration

The provision of public e-services by administration is not only becoming increasingly popular but is beginning to be a standard solution. In an information society, public e-services need to be a priority in the operations of administrative authorities.

C. Gaie
Chief of Engineering and Innovation Department, French Prime Minister Services, 75007 Paris, France

M. Karpiuk (✉)
Chair of Administrative Law and Security Studies, University of Warmia and Mazury, 10-719 Olsztyn, Poland
e-mail: miroslaw.karpiuk@uwm.edu.pl

C. Gaie and M. Mehta (eds.), *Transforming Public Services—Combining Data and Algorithms to Fulfil Citizen's Expectations*, Intelligent Systems Reference Library 252, https://doi.org/10.1007/978-3-031-55575-6_7

The advancement of new technologies has imposed the need to adjust such services to the standards arising from technology development.

The activities of an entity that provides public e-services are not limited to the technical handling of processes and access to a communications network where information is collected, transmitted and stored. Such operations are not purely technical. The provision of these services is not automatic or passive and should cater for society needs that may be fulfilled by electronic means more effectively, efficiently, and cheaply. Such services may reach a wider group of recipients.

Considerable challenges are posed by the legal provisions that hinder e-Administration activities, which should be characterised by innovativeness. Legislators need to keep pace with more and more novel technological solutions aimed at streamlining the e-service sector in which e-Administration is also active. e-Administration should be competitive in this market, and no legal provisions should restrict this competitiveness. Moreover, legislators cannot treat e-administration as superior in relation to other entities functioning in the sphere of e-services [1].

Restrictions of e-Administration activities in the sphere of public services also result in the digital exclusion of a part of society. Therefore, public authorities need to strive for developing digital competencies in society and to ensure widespread internet access. The development of an information society, technological progress, and growing digital competencies among citizens define the status of e-administration in a given state and its place in the e-service sector. The quality and reach of services rendered by electronic means demonstrate the quality of the state and the administration authorities functioning within its structures [2].

eAdministration must be both effective and secure. Given that special emphasis should be placed on cybersecurity. Information and communication (ICT) systems used by public administration authorities must be resilient to the threats that disrupt their operation.

In the digital revolution era, the growing expectations of citizens and other entities, including enterprises, towards the state and the services it renders, constitute a major challenge and an additional impulse for changes to the public sector. Citizens, whose needs are the starting point for any actions taken, should be placed at the centre of the digital transformation occurring in the state and thus in public administration authorities. A shift in the approach to processes, service implementation, and communication with citizens should be based on opportunities arising from the application of digital technologies [3].

The status of e-Administration and the availability of public services provided via e-administration solutions.

7.2 Context of European Union Regulation

This section describes the European digital agenda driven by the European Commission. Figure 7.1 illustrates the timeline to facilitate the understanding:

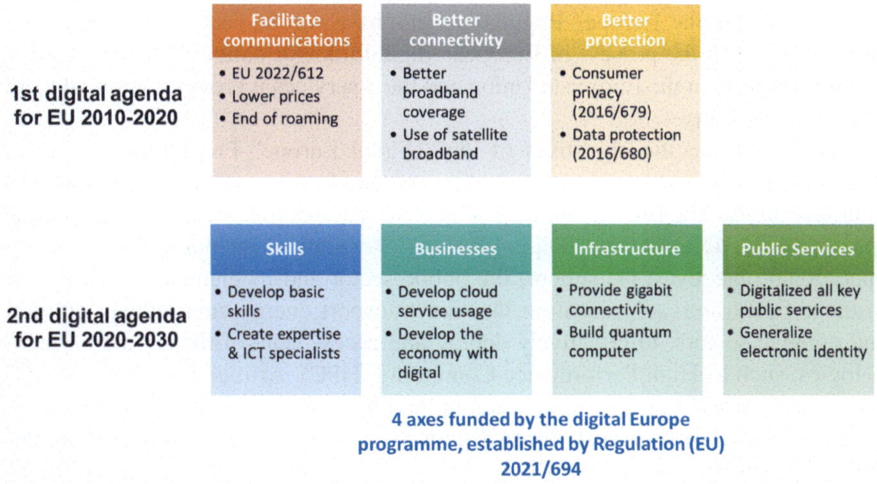

Fig. 7.1 Digital agenda of the European Commission

The European Union has placed emphasis on the status of e-administration and public services provided via e-administration solutions. According to Recital 51 of Regulation (EU) 2021/694 of the European Parliament[1] and of the Council of 29 April 2021 establishing the Digital Europe Programme and repealing Decision (EU) 2015/2240 (Official Journal EU L 166, p. 1), the modernisation of public administrations in European Union Member States is one of the key priorities for the implementation of the Digital Single Market. At the same time, EU legislators highlighted the need to strengthen the transformation of public administrations towards computerisation and to ensure that citizens have easy and secure access to public services.

In Recital 53 of the "Digital Europe" Regulation, it is noted that the interoperability of European public services should concern all levels of administration, from local, regional and national levels, to the European Union level. Interoperability not only allows the removal of barriers to the functioning of an internal market, but it also facilitates cross-border cooperation and the promotion of European standards, which contributes to the emergence of new, and the consolidation of developing, common public services at the European Union level. The interoperability of European public services is understood broadly as a multi-faceted phenomenon spanning technical and legal layers and encompassing policy elements. In practice, a valuable proposal concerning interoperability governance in Europe was proposed by Wimmer et al. in [4]. Another interesting declination is the framework proposed by Campmas et al. outlines the crucial role of interoperability to accelerate the adoption of e-Administration [5].

The general objectives of the "Digital Europe" Programme are to support and accelerate the digital transformation of the European economy, industry and society, and to bring its benefits to public administrations, which is expressly stated in Article

[1] http://data.europa.eu/eli/reg/2021/694/oj.

3 (1) of the "Digital Europe" Regulation. Emphasis on the need to include public administration in the process of European digital transformation demonstrates that e-administration in the European Union and the e-services it provides are vital to an information society.

One of the specific objectives of the "Digital Europe" Programme, referring directly to public e-services, is the deployment and best use of digital capacity and interoperability. The fulfilment of this objective requires the undertaking of actions specified in Article 8 of the "Digital Europe" Regulation, including the following operational objectives: (1) support the public sector and its elements, such as, for instance, health and care, education, taxation, transport, energy, environment, cultural and creative sectors, to effectively deploy and access state-of-the-art digital technologies, such as High Performance Computing (HPC), Artificial Intelligence (AI) and cybersecurity; (2) deploy, operate and maintain trans-European interoperable state-of-the-art digital service infrastructures across the European Union, in complementarity with national and regional actions; (3) support the integration and use of trans-European digital service infrastructures and of European digital standards in the public sector to facilitate cost-efficient implementation; (4) facilitate the development and update of public administrations; (5) offer the public sector easy access to the testing of advanced digital technologies; (6) support the uptake of advanced digital technologies by the public sector; (7) support the design, testing, implementation, deployment and maintenance of interoperable digital solutions, including digital government solutions, for public services at the European Union level; (8) ensure continuous capacity at the European Union level to observe, analyse and adapt to fast-evolving digital trends, and share and mainstream best practices, including in the public e-service sector; (9) support cooperation towards achieving a European ecosystem for trusted data sharing and digital infrastructures; and (10) build up and strengthen the European Digital Innovation Hubs and their network.

In the European Union, there is a notable trend related to the optimum use of interoperable digital tools in the sphere of public services. Emphasis is being placed on developing electronic administrations and their digital capabilities and fostering digital competencies in society while public e-services are gaining popularity. The widespread access to e-services provided by public administration should not be at the expense of quality and security, so it is necessary to incur appropriate financial outlays for modernising the public e-service sector. To illustrate this discussion, the authors recommend to consider the usage of artificial intelligence that was illustrated in [6] and the considerations including cybersecurity [7].

Public services provided by electronic means must be decentralised, and central government authorities should support them. Therefore, administration structures located near citizens need to be equipped with appropriate electronic tools to allow the provision of high-quality services.

According to Recital 18 of Decision (EU) 2022/2481 of the European Parliament and of the Council of 14 December 2022 establishing the Digital Decade Policy Programme 2030 (Official Journal EU L 323, p. 4), further referred to as the Digital Decade Decision, democratic life and key public services depend crucially on digital technologies. Every citizen and business should be able to interact digitally with public administrations, this refers to e-Democracy [8]. The parameters of such interactions include user-centricity and transparency. Given the above, key public services should be fully accessible on a voluntary basis. As stipulated in Recital 19 of the Decision, digital technologies should contribute to achieving society outcomes that are not limited to the digital sphere but also have positive effects on the everyday lives of citizens. Proper digital transformation should go hand-in-hand with improvements regarding democracy, good governance, and more efficient public services.

One of the European Union's digital targets stipulated in Article 4 of the Digital Decade Decision is the digitalisation of public services. The European Union is aware that the provision of services by electronic means is a solution that is likely to facilitate efficient public administration activities which require an innovative mindset, such as the usage of blockchain [9]. It is a prospective approach, focusing on developing an information society whose needs are catered for in the form of public e-services instead of direct contact with a given authority in its office.

Directive (EU) 2016/2102 of the European Parliament and of the Council of 26 October 2016 on the accessibility of the websites and mobile applications of public sector bodies (Official Journal EU L327, p. 1), further referred to as the Web Accessibility Directive—in Recital 1,—expressly indicates that a digital society provides users with new ways of accessing information and services. Public-sector providers of information and services rely increasingly on the Internet in order to produce, collect and provide a wide range of information and services online, which are essential to the public. Therefore, the EU legislator has pointed to the trend of developing public e-services, where administration should be given great significance in respect of the provision of such services. The Directive aims to ensure that the websites and mobile applications of public sector bodies are more accessible based on accessibility requirements across all European Union Member States. To fulfill the accessibility requirements many approaches were detailed such as the harmonization of standards described in [10].

As per Article 4 of the Web Accessibility Directive, the EU legislator obliges the Member States to ensure that public sector bodies take the necessary measures to make their websites and mobile applications more accessible. They should be made perceivable, operable, and understandable. As stipulated in Article 3 (2) of the Directive, "mobile application" means application software designed and developed, by or on behalf of public sector bodies, for use by the general public on mobile devices such as smartphones and tablets. It does not include the software that controls those devices (mobile operating systems) or hardware.

The digital environment in which public administration functions should, on the one hand, guarantee easy access to high-quality public e-services, and ensure user security and high privacy protection standards, on the other hand.

The right to privacy (including privacy in cyberspace) is safeguarded by domestic and international law. Under Article 7 of the Charter of Fundamental Rights of the European Union (Official Journal EU of 2007, C 303, p. 1), everyone has the right to respect their private and family life, home and communications. The right is also guaranteed under Article 8 of the Convention for the Protection of Human Rights and Fundamental Freedoms completed in Rome on 4 November 1950 (Journal of Laws of 1993, No. 61, item 284, as amended), where it is stipulated that everyone has the right to respect for his private and family life, his home and his correspondence. This topic was interestingly tackled by Wu that proposed a benchmark of data protection process for e-Government [11].

As emphasised further in the provision, there shall be no interference by a public authority with the exercise of this right except such as is in accordance with the law and is necessary for a democratic society in the interests of national security, public safety or the economic well-being of the country, for the prevention of disorder or crime, for the protection of health or morals, or the protection of the rights and freedoms of others.

It should be stressed that cybersecurity should be subject to special protection as one of the security domains, especially nowadays, when the fast transmission of information is the foundation of every activity, and digitalisation is a factor contributing to the development of the state and the society, even when it requires actions hindering individual rights and liberties, where necessary, and if the exercise of such rights cannot be reconciled with the assurance of state cybersecurity [12, 13]. The protection of privacy may be restricted in specific cases if it stands in the way of the need to provide cybersecurity.

7.3 Development and Usage of e-Services

For many years, public administration has been taking measures intending to develop e-services. They include, in particular: (1) providing a wide range of public administration services by electronic means, (2) fostering the effectiveness of public administration through interoperable IT solutions, (3) providing the citizens and other entities with data from registers and other public sector information to be further used for the expansion of service portfolios, and (4) supporting the development of e-services with a wide reach and mutual recognition of ICT tools and solutions [14, 15].

The development of e-administration needs to go hand in hand with improving users' digital skills. Digital competencies are a condition for full participation in contemporary societies, as well as a chance to improve the quality of life and to create a proactive approach to novel digital technologies. The objective of digital education is to develop digital competencies enabling citizens to exploit the potential of new technologies in various spheres (including communication with administration authorities) to derive benefits and improve the quality of their lives [16]. Citizens' advanced digital skills constitute a vital factor in building social capital, which also translates into economic growth. In the individual dimension, the effects of digital education include a better alignment of competencies to the labour market, preparation for safe, ethical and skilful use of the opportunities provided by digital technologies, and more widespread use of e-services and cultural, scientific and knowledge-related resources. In the social dimension, the effect of digital education is preparing individuals for participation and communication within the web environment in the spirit of respect for various attitudes and views.

Regarding public administration, in addition to officials' digital competencies, society's computer skills are also vital because efficient public e-structures will only be possible if citizens are not only willing but also able to handle their affairs by electronic means, as this both facilitates faster decision-making in residents' matters and allows the reduction of administration costs. Opening up to digital technologies is favourable to improving operational efficiency, the quality of services being provided, building a positive image of administration in society and developing trust in its authorities [17, 18]. The application of ICT systems provides space for improving the efficiency of the tasks being implemented and extending their scope. The modernisation of public administration (including local governments) based on state-of-the-art communication tools contributes to enhancing the quality of customer service and relations between public offices and citizens.

Using the opportunities offered by digital technologies warrants greater transparency of public structures' operations, thus facilitating better social control [19, 20]. The objective of the digital transformation of these structures is not only to streamline their functions but also to reach the broadest possible group of recipients using their e-services.

The development of e-services (including services in the public sphere) is related to the introduction of modern communication tools. Their deployment might also result in the emergence of threats. Yet, we need to look forward to the future with optimism, as the aim of e-services is, most of all, to ensure better communication and to speed up the procedures for handling various matters [21, 22]. Notwithstanding the essence of a specific service, we are due to witness the continuous growth of new solutions in the scope of public e-services, which will undoubtedly limit, and increasingly phase out, traditional contacts with public offices.

Electronic delivery services belong to the set of tools being used in e-administration. As per Article 4 of the Electronic Delivery Service Act of 18 November 2020 (consolidated text, Journal of Laws of 2023, item 285, as amended), a public authority shall deliver correspondence that requires acknowledgement of receipt or receipt via a public electronic registered delivery service to the electronic delivery address entered in the database of electronic addresses. Under Article 3 (36) of Regulation (EU) No. 910/2014 of the European Parliament and of the Council of 23 July 2014 on electronic identification and trust services for electronic transactions in the internal market and repealing Directive 1999/93/EC (Official Journal EU L 257, p. 73), an electronic registered delivery service means a service that makes it possible to transmit data between third parties by electronic means and provides evidence relating to the handling of the transmitted data, including proof of sending and receiving data, and that which protects transmitted data against the risk of loss, theft, damage or any unauthorised alterations [23].

The essence of the principle of electronic delivery primacy is entailed in the arrangement of correspondence exchange via a public electronic registered delivery service to an electronic delivery address (entered in the database of electronic addresses). This delivery of correspondence by electronic means is supplemented by a hybrid public service which consists of a combination of electronic delivery and traditional handing-over of a letter personally to the addressee. Hybrid public

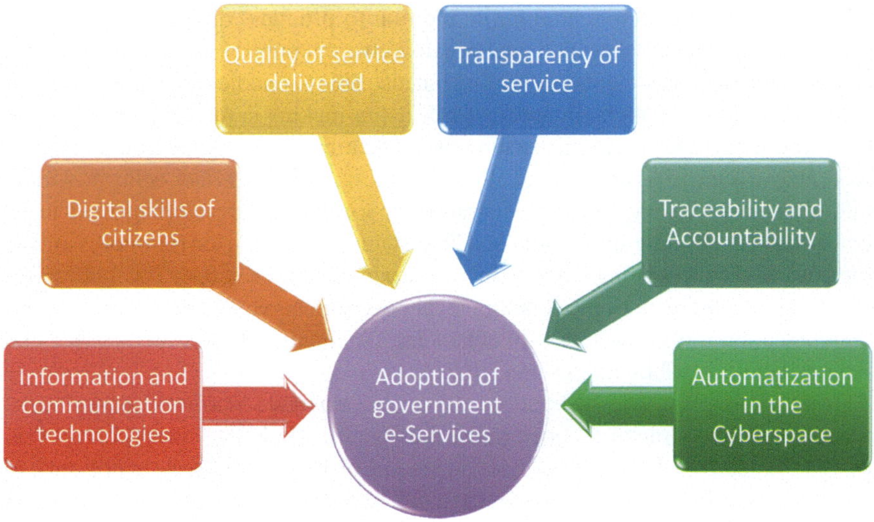

Fig. 7.2 Factors that contribute to the adoption of digital public services

services are addressed to individuals affected by digital exclusion and those not ready to resign from the traditional delivery method for reasons other than digital exclusion.

Public administration authorities provide e-services in cyberspace. Cyberspace is defined as a global network composed of interconnected ICT systems built of devices which allow the generation, processing and exchange of information in an automated way between devices or users, acting intentionally and purposefully [24, 25]. Figure 7.2 illustrates the combination of factors that contribute to the adoption of e-services delivered by public administration.

7.4 Cybersecurity of Public e-Services

The efficient functioning of the state requires not only the performance of tasks with the use of cyberspace but also the assurance of cybersecurity. However, the protection of cyberspace must be continuous, not only in times of crisis or conflict (although it is particularly important in such circumstances) but also in a situation where the state uninterruptedly performs its tasks [26]. The intensity of the protection may vary depending on the cyber threats that have arisen or might occur.

Cybersecurity is one of the specific objectives of the European Union's "Digital Europe" programme. Within this sphere, the following operations are performed: (1) supporting the building-up and procurement of advanced cybersecurity equipment, tools and data infrastructures, together with Member States, in order to achieve a high common level of cybersecurity; (2) supporting the building-up and best use of knowledge, capacity and skills related to cybersecurity, and the sharing and mainstreaming of best practices; (3) ensuring the deployment of effective state-of-the-art cybersecurity solutions (including in public administration); (4) reinforcing capabilities within Member States to help them uptake cybersecurity best practices; (5) improving resilience against cyber-attacks, contributing towards increasing risk-awareness and knowledge of cybersecurity processes, and supporting public organisations in achieving basic levels of cybersecurity, as well as (6) enhancing cooperation between the civil and defence spheres concerning dual-use projects, services, competences and applications in cybersecurity. The specific objective and the actions to be taken, respecting its fulfilment, are set out in Article 6 of the "Digital Europe" Regulation [27].

The Union's Digital Decade Programme assumes that the objectives of the European Union are to improve resilience against cyber-attacks, to contribute to increased risk awareness and knowledge of cybersecurity processes, and to increase the efforts of public organisations to achieve at least basic levels of cybersecurity. The objectives can be derived directly from Article 3 (1) of the "Digital Decade" Decision.

As stated in Recital 3 of Directive (EU) 2022/2555 of the European Parliament and of the Council of 14 December 2022 on measures for a high common level of cybersecurity across the Union, amending Regulation (EU) No. 910/2014 and Directive (EU) 2018/1972, and repealing Directive (EU) 2016/1148 (Official Journal EU L 333, p.80), network and information systems have developed into a central feature of everyday life with the speedy digital transformation, including in cross-border exchanges. That development has led to an expansion of the cyber threat landscape, bringing about new challenges, which require responses that are coordinated, innovative and adapted to a given situation in all European Union Member States. The number, sophistication, frequency and impact of incidents are increasing and present a major threat to the functioning of network and information systems (including those used by public administration). As a result, incidents can impede the pursuit of economic activities in the internal market, generate financial loss, undermine user confidence, and cause major damage to the European Union's economy and society. Cybersecurity preparedness and effectiveness are, therefore, now more essential than ever to the proper functioning of the EU internal market.

Cybersecurity is defined under Article 2 (1) of the Regulation (EU) 2019/881 of the European Parliament and of the Council of 17 April 2019 on ENISA (the European Union Agency for Cybersecurity) and information and communications technology cybersecurity certification and repealing Regulation (EU) No. 526/2013 (Official Journal EU L 151, p. 15), further referred to as the ENISA Regulation, and it means the activities necessary to protect network and information systems, the users of such systems, and other persons affected by cyber threats. Under Article 2 (4) of the National Cybersecurity System Act of 5 July 2018 (consolidated text, Journal of Laws of 2022, item 1863, as amended), hereinafter the Cybersecurity System Act, cybersecurity is understood as the ability of information systems to resist action that compromises the confidentiality, integrity, availability, and authenticity of processed data or the related services offered by those information systems. Cybersecurity is one of the security system components which comprises the protection of information systems against threats [28].

As laid down in Article 2 (8) of the ENISA Regulation, a cyber threat is defined as any potential circumstance, event or action that could damage, disrupt or otherwise adversely impact network and information systems, the users of such systems and other persons.

The European Union Agency for Cybersecurity (ENISA) supports capacity-building and preparedness across the European Union by assisting the Union institutions, bodies, offices and agencies, as well as Member States and public and private stakeholders, to increase the protection of their network and information systems, to develop and improve cyber resilience and response capacities, and to develop skills and competencies in the field of cybersecurity. The objective of the Agency is stipulated in Article 4 (3) of the ENISA Regulation. It constitutes institutional support for the public sector, including administration, in the sphere of developing a system of protection against cyber threats, including where public e-services are provided [29].

An incident, as an event that has or might have an adverse effect on cybersecurity, may also happen in a public entity providing e-services. Under Article 2 (9) of the Cybersecurity System Act, an incident in a public entity means an incident that results in or might result in the deterioration of quality or disruption of public tasks performed by a public entity.

As stipulated in Article 21 (1) of the Cybersecurity System Act, a public entity that performs public tasks relying on information systems is obliged to appoint a person responsible for maintaining contacts with national cybersecurity system entities. Thus, the legislator has imposed an obligation to entrust this function to one person, having found that it would be sufficient, and not considering that the specific nature and intensity of tasks relying on an information system might require an entire team for such contacts [30].

As per Article 22 of the Cybersecurity System Act, a public entity performing public tasks relying on information systems (including the provision of e-services) must: (1) ensure incident handling; (2) immediately notify a competent CSIRT about incidents, no later than within 24 h of detection; (3) ensure incident handling in collaboration with the competent CSIRT, by providing the required data, including personal data; (4) provide persons for whom a public task is performed with access to knowledge that allows the understanding of cybersecurity threats and the use of appropriate safeguarding measures, in particular by publishing relevant information on its website; and (5) transmit to the competent CSIRT the personal data of a person responsible for contacts with national cybersecurity system entities [31].

The main objectives of the Polish Cybersecurity Strategy are to improve the level of resilience to cyber threats, to increase the level of information protection in the public, military and private sectors, and to promote knowledge and good practices to help citizens better protect their information. The specific objectives include increasing the level of resilience of information systems of the public administration and achieving the capacity to prevent and respond to incidents effectively [32, 33].

Professional staff that perform cybersecurity tasks and have appropriate knowledge, proper skills and qualifications guarantee the quality of actions safeguarding cyberspace against attacks and thus contribute to optimising activities in this sphere and help minimise disruptions in this area [26]. Cybersecurity specialists are continually sought by international corporations and Polish enterprises representing various industries, critical infrastructure operators or special services. In this cross-sectoral competition, the public sector is often in a difficult position related to limited financial capacity [32]. To retain highly skilled employees in public administration, the application of instruments supporting their activities in respect of cybersecurity protection needs to be accompanied by other actions aimed at approximating the salaries of these staff members to the levels they would be able to reach in the private sector.

Both public administration employees and the citizens whose needs are catered for with the use of ICT systems should have digital competencies, while there should also be a will to turn to e-services among the citizens, as they are the ones for whom the services are rendered.

As regards the cybersecurity of public e-services, it is also worth discussing the planning component. In their planning documents, public administration needs to take cybersecurity into account as a vital element that is to ensure the efficient performance of public tasks via ICT systems which require protection against cyber threats. Planning, including the one related to activities in cyberspace, facilitates coordinated measures allowing a proper, timely and balanced performance of tasks assigned to public administration in a well-organised and uninterrupted manner [34].

7.5 Conclusions and Perspectives

To ensure that official matters are handled conveniently, electronic communication is of paramount importance not only between citizens and administration but also between specific public administration authorities. The access to public e-services allows officials to perform their operations via information systems instead of customers personally appearing before public administration authorities, as well as to streamline and speed up customer service for those who rely on electronic tools. This all contributes to building social trust in public administration and the initiatives undertaken.

Advanced digital technologies are an inherent part of our professional and private reality, and they are gradually becoming a vital element of the public sector, while both administration authorities and customers need to adopt an open-minded approach to these transformations. The modernisation of public administration and the services it provides with the use of digital tools is of fundamental importance to the efficient functioning of the state intending to reduce the burdens placed on the society and other entities (including enterprises). It also allows the reduction of the overall costs of administrative operations and better social control over these tasks due to the increase in their transparency (broader access to information about public administration activities). To keep pace with the changes resulting from the development of technology that is used for providing e-services, public administration not only needs to take proper care of the ongoing operations of information systems but also upgrade them, focus on expanding the knowledge of its employees and their skills, and take up administrative measures to streamline the functioning of e-administration.

The continuous modernisation of public administration, focused on the application of digital technologies, should be the principal goal of state operations in the sphere of public e-services. Such modernisation should be directed towards the need to boost the efficiency of the state by improving the quality of relationships between administration, citizens, and other stakeholders. The use of the opportunities arising from new digital technologies facilitates greater transparency and effectiveness of activities in the public sphere. The primary beneficiaries of actions taken by the state for digital transformation are citizens who gain access to high-quality advanced e-services provided by public administration. Only legislative and organisational changes that consider technological progress will allow the elimination of ineffective and excessively formalised procedures and contribute to minimising the burdens imposed on the citizens. Digital tools used by public administration must be based on state-of-the-art solutions that facilitate their comprehensive and multiple application. Such legal, organisational, and technological bases ensure a more efficient e-administration management system, both in the context of internal activities of individual offices and the handling of matters where several entities are engaged.

Emphasis should also be placed on the security of information systems (cybersecurity) through which public services are rendered by preserving citizens' privacy rights [35]. Cybersecurity, as a type of security, should be perceived as a common interest which needs to be protected by law and, for that reason, it gains special importance and is vital from the perspective of the normal functioning of the state and also from the point of view of meeting information society's needs [36].

Common interest constitutes the foundation of public order and the democratic system of a state of law, thus becoming an objective pursued by public authorities. Common interest is a multidimensional category that is not easy to define, which does not mean that public entities may waive responsibility for its pursuance.

The current pace of information technology development provides reasons to believe that cybernetic threats may affect a growing proportion of our population and exert an increasing impact on the functioning of states. Future research will detail various methods and approach to protect government services from these cyber threats.

References

1. Hofmann, H., & Türk, A. (Eds.). (2009). *Legal challenges in eu administrative law: Towards an integrated administration*. Edward Elgar Publishing.
2. Kim, D., & Grant, G. (2010). e-Government maturity model using the capability maturity model integration. *Journal of Systems and Information Technology, 12*(3), 230–244. https://doi.org/10.1108/13287261011070858

3. Integrated State Computerisation Programme. (2019), p. 15. https://digital-skills-jobs.europa.eu/en/actions/national-initiatives/national-strategies/poland-integrated-state-computerisation-programme
4. Wimmer, M. A., Boneva, R., & di Giacomo, D. (2018). Interoperability governance: a definition and insights from case studies in Europe. In *Proceedings of the 19th annual international conference on digital government research: Governance in the data age (dg.o '18).* Association for Computing Machinery, New York, NY, USA, Article 14, 1–11. https://doi.org/10.1145/3209281.3209306
5. Campmas, A., Iacob, N., & Simonelli, F. (2022). How can interoperability stimulate the use of digital public services? An analysis of national interoperability frameworks and e-Government in the European Union. *Data & Policy, 4*, e19. https://doi.org/10.1017/dap.2022.11
6. Henman, P. (2020). Improving public services using artificial intelligence: Possibilities, pitfalls, governance. *Asia Pacific Journal of Public Administration, 42*(4), 209–221. https://doi.org/10.1080/23276665.2020.1816188
7. Krishna, B., & M.P., S. (2021). Examining the relationship between e-Government development, nation's cyber-security commitment, business usage and economic prosperity: A cross-country analysis. *Information and Computer Security, 29*(5), 737–760. https://doi.org/10.1108/ICS-12-2020-0205
8. Abu-Shanab, E. (2015). E-democracy: The fruit of e-Government. *International Journal of Technology and Globalisation., 8*, 15–28. https://doi.org/10.1504/IJTG.2015.077873
9. Kassen, M. (2022). Blockchain and e-Government innovation: Automation of public information processes. *Information Systems, 103*, C. https://doi.org/10.1016/j.is.2021.101862
10. Moreno, L., & Martinez, P. (2019). The harmonization of accessibility standards for public policies. *Computer, 52*(7), 57–66. https://doi.org/10.1109/MC.2018.2888762
11. Wu, Y. (2014). Protecting personal data in e-Government: A cross-country study. *Government Information Quarterly, 31*(1), 150–159. ISSN 0740-624X. https://doi.org/10.1016/j.giq.2013.07.003
12. Hoffman, I., & Karpiuk, M. (2022). E-administration in polish and Hungarian municipalities—A comparative analysis of the regulatory issues. *Lex localis—Journal of Local Self-Governments, 20*(3), 617–640. https://doi.org/10.4335/20.3.617-640(2022
13. Karpiuk, M. (2022). Recognizing an entity as an operator of essential services and providing cybersecurity at the national level. *"Prawo i Więź" 42*(4). https://doi.org/10.36128/priw.vi42.524
14. Romaniuk, P. (2016). Electronic administration as a factor in the efficient management development and communication of public administration. *International Journal of Contemporary Management, 2016*, 79–94.
15. Romaniuk, P. (2022). Legal aspects connected with the preparation of risk management mechanisms in public administration. *Cybersecurity and Law, 8*(2), 84–94. https://doi.org/10.35467/cal/157126
16. Morte-Nadal, T., & Esteban-Navarro, M. A. (2022). Digital competences for improving digital inclusion in e-Government services: A mixed-methods systematic review protocol. *International Journal of Qualitative Methods, 21*. https://doi.org/10.1177/16094069211070935
17. Mensah, I. K. (2020). Impact of government capacity and e-Government performance on the adoption of e-Government services. *International Journal of Public Administration, 43*(4), 303–311. https://doi.org/10.1080/01900692.2019.1628059
18. Abdulkareem, A. K., & Mohd Ramli, R. (2022). Does trust in e-Government influence the performance of e-Government? An integration of information system success model and public value theory. *Transforming Government: People, Process and Policy, 16*(1), 1–17. https://doi.org/10.1108/TG-01-2021-0001
19. Arie Halachmi & Dorothea Greiling. (2013). Transparency, e-Government, and accountability. *Public Performance & Management Review, 36*(4), 562–584. https://doi.org/10.2753/PMR1530-9576360404
20. Sabani, A. (2021). Investigating the influence of transparency on the adoption of e-Government in Indonesia. *Journal of Science and Technology Policy Management, 12*(2), 236–255. https://doi.org/10.1108/JSTPM-03-2020-0046

21. Homburg, V. (2018). ICT, e-Government and E-Governance: Bits & Bytes for Public Administration. In E. Ongaro, S. Van Thiel (eds) *The Palgrave handbook of public administration and management in Europe*. Palgrave Macmillan. https://doi.org/10.1057/978-1-137-55269-3_18
22. Clifton, J., Díaz Fuentes, D., & Llamosas García, G. (2020). *ICT-enabled Co-production of public services: barriers and enablers. A systematic review* (pp. 25–48). https://doi.org/10.3233/IP-190122
23. Sharma, S., Kar, A. K., & Gupta, M. P. (2022). Unpacking digital accountability: Ensuring efficient and answerable e-governance service delivery. In *Proceedings of the 14th international conference on theory and practice of electronic governance (ICEGOV '21)*. Association for Computing Machinery, New York, NY, USA (pp. 260–269). https://doi.org/10.1145/3494193.3494229
24. Chałubińska-Jentkiewicz, K., Karpiuk, M., & Kostrubiec, J. (2021). The legal status of public entities in the field of cybersecurity in Poland. *Lex localis. Maribor: Institute for Local Self-Government.* https://doi.org/10.4335/2021.5
25. Hoffman, I., & Karpiuk, M. (2022). The local self-government's place in the cybersecurity domain. Examples of Poland and Hungary. *Cybersecurity and Law, 7*(1), 171–190. https://doi.org/10.35467/cal/151826
26. Bencsik, A., & Karpiuk, M. (2023). Cybersecurity in Hungary and Poland. Military aspects. *Cybersecurity and Law, 9*(1), 82–94. https://doi.org/10.35467/cal/169302
27. Fuster, G. G., & Jasmontaite, L. (2020). Cybersecurity regulation in the European Union: The digital, the critical and fundamental rights. In M. Christen, B. Gordijn, M. Loi (Eds.), *The Ethics of Cybersecurity. The International Library of Ethics, Law and Technology* (vol. 21). Springer. https://doi.org/10.1007/978-3-030-29053-5_5
28. Czuryk, M. (2019). Supporting the development of telecommunications services and networks through local and regional government bodies, and cybersecurity. *Cybersecurity and Law, 2*. https://doi.org/10.35467/cal/133839
29. Markopoulou, D., Papakonstantinou, V., & de Hert, P. (2019). The new EU cybersecurity framework: The NIS Directive, ENISA's role and the general data protection regulation. *Computer Law & Security Review, 35*(6), 105336. ISSN 0267-3649. https://doi.org/10.1016/j.clsr.2019.06.007
30. Karpiuk, M. (2020). The obligations of public entities within the national cybersecurity system. *Cybersecurity and Law, 2*. https://doi.org/10.35467/cal/133971
31. Ponsard, C., & Darimont, R. (2020). Regulation and security modelling of essential services in network of information systems. iStar. https://ceur-ws.org/Vol-2641/paper_08.pdf
32. Cybersecurity Strategy of the Republic of Poland. https://www.gov.pl/attachment/6a4aafc6-e339-4cd5-a8e6-cd47257f02d8
33. Kitler, W. (2022). The cybersecurity strategy of the Republic of Poland. In K. Chałubińska-Jentkiewicz, F. Radoniewicz & T. Zieliński (Eds.), *Cybersecurity in Poland*. Springer. https://doi.org/10.1007/978-3-030-78551-2_9
34. Karpiuk, M. (2021). Cybersecurity as an element in the planning activities of public administration. *Cybersecurity and Law, 1*. https://doi.org/10.35467/cal/142179
35. Gaie, C. (2021). Optimise data exchange between government administrations while ensuring privacy rights. *International Journal of Business Intelligence and Systems Engineering, 1*(4), 347–357. https://doi.org/10.1504/IJBISE.2021.122786
36. Czuryk, M. (2021) Cybersecurity as a premise to introduce a state of exception. *Cybersecurity and Law, 2*. https://doi.org/10.35467/cal/146466

Chapter 8
Modernizing the Processes for Human Resources Through Artificial Intelligence to Optimize a Variety of e-Government Responsibilities

Christophe Gaie and Markus Mueck

Abstract This chapter tackles the challenge of modernizing the processes for Human Resources of government services. Precisely, it addresses a variety of e-Government responsibilities such as identifying suitable candidates for an open position, advising employees in their career, or proposing a matching between opportunities and available employees. The authors describe a novel architecture that combines successive operations: 1- gathering key information sources, 2- introducing a novel metric that allows to evaluate and compare algorithms, and 3- implementation of a matching through different algorithms. A particular attention is given to the matching that may rely either on classical methods that simply associate skills and experience of candidates and positions; or implement Artificial Intelligence, which carries out an in-depth study combining other information (for example: job location and exposure, expectations and achievements of candidates). Finally, the authors detail how the proposed solution complies with privacy rights' regulation (especially General Data Protection Regulation and Network and Information Systems Directive).

Keywords e-Government · Data science · Data analytics · Artificial intelligence · Human resources · Data protection · Data regulation

C. Gaie (✉)
Chief of Department of Engineering and Innovation, French Prime Minister Services, 75007 Paris, France
e-mail: christophe.gaie@gmail.com

M. Mueck
ETSI Board, 82008 Unterhaching, Germany
e-mail: markus.muck@gmail.com

C. Gaie and M. Mehta (eds.), *Transforming Public Services—Combining Data and Algorithms to Fulfil Citizen's Expectations*, Intelligent Systems Reference Library 252, https://doi.org/10.1007/978-3-031-55575-6_8

8.1 Introduction

Government administrations began to acquire computers and employ them to facilitate the work of civil servants since the 1950s. Then, government states pursued a new objective in the 1970s in order to increase the efficiency of public services through digitalization as indicated in [1, 2]. The initial efforts were dedicated to improve the internal procedures of government services such as replacing paper records by database which contributes to reduce duplicates, avoid fraud, facilitate mobility, etc.

Then, there was a revolution of the Information and Communication Technology (ICT) with the emergence and growth of World Wide Web in the 1990s. This offered the opportunity to offer public services to citizens in a fully digitalized manner. People were required to achieve themselves (through the Internet) actions that contribute to government procedures: report their personal income, require grant subsidies, or schedule an appointment with a public service [3, 4].

The last few years have seen the emergence of e-Government that primarily aims to optimize existing government services such as healthcare organization [5, 6], agriculture efficiency [7] or fraud detection [6, 8]. The introduction of machine learning was a major step to provide better services to citizens. The effort and actions held by researchers as well as IT civil servants largely contributed to accelerate the transformation from old paper-based procedures towards digital services [9].

While this evolution is valuable and necessary to improve public services, the modernization henceforth enters in a new phase. As a matter of fact, the amount of data available and exchanged makes it possible to automatize public services. For example, it is possible to grant a social subsidy without any intervention for a citizen or to withhold tax automatically. We believe this new paradigm should be entitled as "**A-Government**" (Automated Government). Figure 8.1 illustrates the timeline of e-Government evolution:

To enable the automation of public services, the authors suggest investigating the prediction perspective. Indeed, whenever it is possible to anticipate behaviors of citizens, government services may be optimized to offer them a better experience. For example, prediction may offer the opportunity to ensure that they comply with the fiscal legislation or the police law, as well as to provide them better public services in various domains: culture, education, justice, Human Resources, etc.

The current chapter is dedicated to the modernization of the processes for Human Resources through Artificial Intelligence (AI) to optimize a variety of e-Government responsibilities such as described in Fig. 8.2.

The chapter is organized as follows. First, Sect. 8.2 proposes a comprehensive review of from the emergence of AI in e-Government to innovative approaches to modernize The processes for Human Resources Then, Sect. 8.3 describes a framework to optimize these processes in the context of e-Government applications. This framework optimizes the recruitment of candidates as well as career management. Thereafter, Section 8.4 proposes a method to ensure trust in the system. Finally, Sect. 8.5 provides a conclusion and describes future directions for this work.

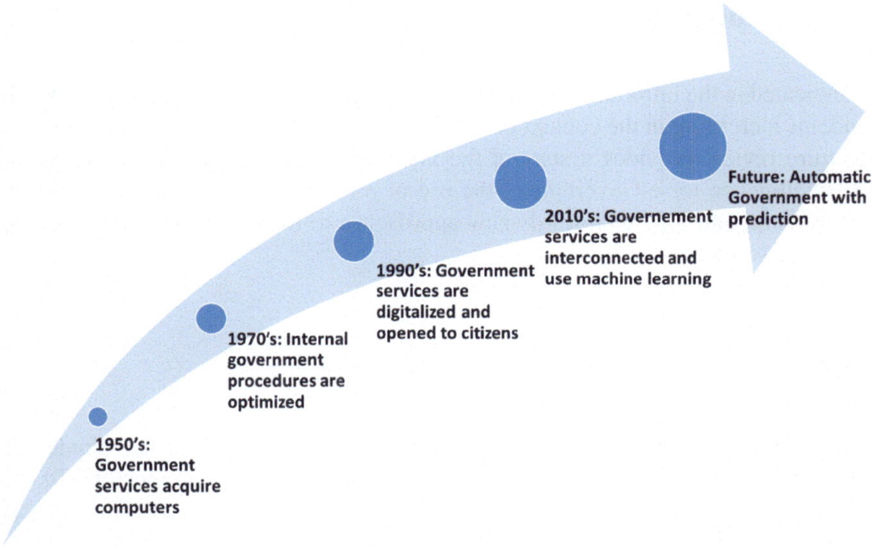

Fig. 8.1 The evolution of e-Government until the present time

Fig. 8.2 The complementary dimensions to enhance the processes for human resources

8.2 Literature Review

As indicated in the introduction, the usage of technology (especially artificial intelligence) is increasing in the context of public services and procedures. The following literature review provides a state-of-the art concerning the application of Artificial Intelligence for e-Government, the use of prediction or personalization for e-Government, and a focus towards new approaches to modernize the processes for Human Resources.

8.2.1 Artificial Intelligence for e-Government

After several years of growth, the use of Artificial Intelligence for e-Government is now widespread and has a large variety of applications for the public sector [10]. As a matter of fact, Wirtz et al. describe the link between the branch of AI employed (Knowledge Management, Automation System, Digital Assistant, etc.), the potential value (sort and share knowledge, automate standard tasks, answer to basic questions, etc.) and the concrete use case with appropriate research paper as a reference. The paper also outlines and describes four major challenges to apply AI in the public sector: technology implementation, law and regulation, ethics, and society adjustment. This paper provides meaningful foundations for implementing AI for government services and can be optimized through a concrete implementation that will necessarily require adjusting the model to local particularities (law and regulation, existing information system, tradeoff between challenges, etc.).

Artificial Intelligence also applies through various approaches such as Machine Learning, Deep Learning, Natural Languages Processing, Context Awareness, or Data Security and Privacy as detailed by Al-Mushayt [11]. The author provides very useful tools by proposing 1—A framework to facilitate the combination of e-Government information resources, 2—Deep learning models to provide automatized e-Government services, and 3—A new architecture that supports the implementation of new AI government services. While these proposals are really promising, their concrete implementation is not achieved and described.

Another valuable approach consists of combining data obtained from social media or other sources through big data or AI algorithms in order to improve public policies. To achieve this objective, Kaya proposed a conceptual model, entitled Engage, in [12] that aims to predict the citizens' decisions based on the opinion they express on social media. This model is very simple and can be adapted to various government processes. A drawback of this flexibility relies on the necessity of adapting it in a precise context.

A more sophisticated framework was proposed by Al-Besher and Kumar to enhance e-Government through the usage of Internet of Things and Artificial Intelligence [13]. The model proposed is self-standing, as it does not depend on technical platforms and organizational structure. It considers the rules and principles from IoT and AI, as well as regulation rules. Thus, it may be applied to various public domains such as security and safety, environmental pollution, e-Health, e-Agriculture, e-Bill payment, etc. While the model is promising, the paper does not detail how to implement it operationally.

8.2.2 Prediction and Personalization for e-Government

The use of Artificial Intelligence should contribute to improve the predictability of citizens' behavior and offer them personalized services. The relation between government services and the population is continuously increasing with the digitalization of public services and there is an increasing number of contacts initiated by citizens. However they were originally dedicated to obtain information or conform to a legal obligation [14]. As a result, there is a vast interest in developing new qualitative approaches to improve the quality of public services and their perception by citizens. To the authors' point of view, this perspective is not detailed in the publication.

An interesting approach to optimize the adoption of e-Government is to follow its acceptance and usage by citizens. Many researchers employed the Unified Theory of Acceptance and Use of Technology (UTAUT) by Venkatesh, Morris to this end. For example, Sanmukhiya introduced a Trust parameter to predict the adoption of e-Government in Mauritius [15]. The author identified multiple factors that contribute to the adoption of e-Government that depend from citizen themselves (knowledge, self-confidence, IT skills, etc.) as well as the quality of the service offered (usability, understandability, confidentiality). This approach is valuable but does not detail how to achieve this adoption of public services.

A valuable approach to customize public services consists in associating citizens with the construction process of public services [16]. This collaboration favors the development of services that take effectively into account citizens' needs. It may be implemented by using the e-Government Groupware framework described in previous publications. Anthopoulos et al. outlined that the personalization of public services increases the adoption of public services and their usage. While the proposal is promising, the implementation of a pilot project is required to validate the orientations.

Beyond the prediction on the usage of public services, there is much research that tackles the concern of their personalization. For instance, Al Hassan et al. outline the importance of developing intelligent public services in the domain of tourism to facilitate the development of their economy [17]. As a matter of fact, providing personalized information optimizes tourists' web browsing by displaying them to appropriate travel destinations. The research details an ontology-based personalized system framework. However, the concrete implementation is not detailed.

It is worth observing that data mining techniques offer the opportunity to profile the behavior of public services users. In [18], Mostafa and El Masry proved the interest of taking advantage of known information (sex, age, educational level, perceived usefulness, ease of use, compatibility, etc.) to predict the usage of public services using three artificial neural network models and three machine learning techniques. The results indicate that perceived usefulness and ease of use largely contribute to e-Government adoption. This information plays a major role in designing new public services and facilitate their adoption by citizens. However, there is no proposal on how to optimize public services from these observations.

The growing maturity of digitalized public services offers the opportunity to exploit data to design new personalized services. In [19], Maksimova et al. outline the existence of a cluster of individuals that either use a limited or many services. The authors take advantage of the data obtained from the X-Road program of Estonia to analyze the behavior of citizens and imagine new public services. While the usage of data reveals crucial to enhance public services, the authors do not explain how to optimize the interactions between citizens and government organizations.

8.2.3 New Approaches to Modernize the Processes for Human Resources

Artificial Intelligence is a well-established notion that was already mentioned by John MacCarthy in [20]. However, its usage in the human resource management process is rather new and follows the digitalization of HR Information Systems (HRIS).

A recent literature review provided by Votto et al. establishes six categories to tackle the domain: employee performance, discipline management, recruitment, salary and benefits, best practices, and training [21]. The paper interestingly describes the HR tasks simplified by AI for the defined categories. For instance, the recruitment process may be optimized through the automated analysis of positive action, the effort to reduce discrimination, or the screening and evaluation of candidates. The research outlines that three domains are more addressed than others (best practices, recruitment, salary, and benefits). Similarly, some AI methods are more represented in the literature (machine learning and recommendation engines).

Recently, a relevant analysis of the existing literature was proposed by Vrontis et al. [22]. The research underlines that employing artificial intelligence, robotics and other advanced technologies within human resource management is a new field of research. The research confirms that AI improves decision-making and practices of the HR domain. It also points out the importance of ensuring the acceptability of AI in this domain, since it modifies the decision processes and raises new ethical challenges. The researchers insist on the need to accompany the changes since the modernization will replace existing jobs and require focusing HR managers on the highest value tasks, such as explaining the decisions made.

To implement Artificial Intelligence in the human resource management (HRM) domain, Pan et al. proposed in [23] to combine the technology, organization, and environment (TOE) model described by Baker in [24] with the transaction cost theory. The objective pursued is to identify the factors that contribute positively or negatively to the usage of AI in the recruitment process. The results show that the most AI is perceived as complex, the least it is adopted; on the contrary, the better it is mastered by IT teams, the better it is embedded in the process. While this approach is promising, it does not describe how to implement AI in a government organization.

A valuable implementation of AI to facilitate HRM was proposed by Gong et al. in the context of worker recruitment [25]. First, the authors identify the important information to extract from the resume of candidates and structure them in a model dedicated to further usage. Then, the researchers estimate the potential salary demands of applicants using the Back Propagation Neural Network (BPNN). The results indicate that the algorithm converges in less than 200 iterations and offers valuable information in the recruitment process. The proposal is promising, but there is no concrete feedback on its utilization in the processes for HR.

To highlight the importance of personalized HRM, Huang et al. propose a conceptual model that establishes a link between the implementation of individual approaches using AI and the financial performance of organizations [26]. As a matter of fact, the optimization of the processes for HR may contribute to improve individual performance by upgrading worker skills, job adequacy, or employee motivation. This contributes to a better productivity of the organization workforce and as a consequence to a better performance of the organization. The authors also provide a useful table that links functions of HRM to personalized HRM and are finally declined in potential HR analytics and AI applications. While this description is valuable the authors do not detail how to implement it concretely.

8.2.4 Summary of the Literature Review

To conclude this section, the following table summarizes the contributions and limits of the papers (Table 8.1).

8.3 A Framework to Improve the Processes for Human Resources

In the following paragraphs, we illustrate a possible initialization of a matching that identifies the best candidates for available jobs within a government administration. The proposed algorithm is inspired by the functioning of the application VINCI that is employed by the French government to propose jobs to top managers of the French public administration. The algorithm is based on a combination of skills, experience in the domain, and job seniority. Figures 8.3 and 8.4 illustrate the proposed framework and its matching (Table 8.1).

First, we propose to compute a score that indicates the adequacy between the skills detained by a candidate and the required skills to achieve the tasks required by the job. While we maintain the measure of the skill generic, such that it can be

Table 8.1 Contributions and limits of the different papers cited in the literature review

References	Contribution	Limits
[10]	Provides meaningful foundations for implementing AI for government services	Does not provide a concrete implementation that will necessarily require adjusting the model to local particularities
[11]	Proposes very useful tools to achieve new AI government services	Does not detail a concrete implementation
[12]	Describes a very simple implementation that can be adapted to various government processes	The flexibility induces to adapt the solution in any precise context
[13]	May be applied to various public domains	Does not detail how to implement it operationally
[14]	Provides a state-of-the-art that underlines new approaches to improve public services	Should be updated with current approach
[15]	Identifies multiple factors that contribute to the adoption of e-Government	Does not detail how to achieve this adoption of public services
[16]	Outlines that personalizing public services increases the adoption of public services and their usage	The implementation of pilot project is required to validate the orientations
[17]	Underlines that personalized information optimizes tourists' web browsing	Details an ontology-based framework but does not implement it
[18]	Indicates that perceived usefulness and ease of use largely contribute to e-Government adoption	There is no proposal on how to optimize public services from these observations
[19]	Uses data obtained from the X-Road program of Estonia to analyze the behavior of citizens	Does not indicate how to optimize the interactions between citizens and government organizations
[21]	Highlights the HR tasks simplified by AI in the literature	The analysis is performed on a small sample of 33 research articles
[22]	Confirms that AI improves decision-making and practices of the HR domain and points out the importance of ensuring the acceptability of AI in this domain	The state-of-the-art is not exhaustive and could be further extended (notably by using other keywords)
[23]	Outlines that simplicity contributes to AI adoption	Does not describe how to implement AI in a government organization
[25]	The algorithm converges rapidly and offers valuable information in the recruitment process	There is no concrete feedback on its utilization in the processes for HR
[26]	Highlights the importance of personalized HRM and proposes a useful conceptual model	Does not detail how to implement it concretely

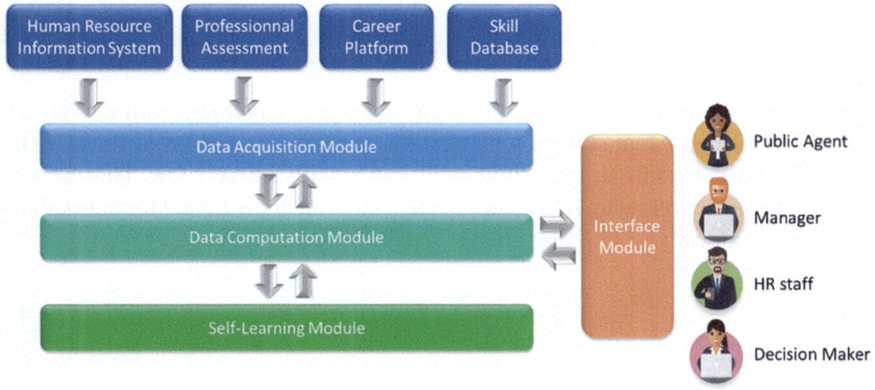

Fig. 8.3 Description of the proposed framework architecture

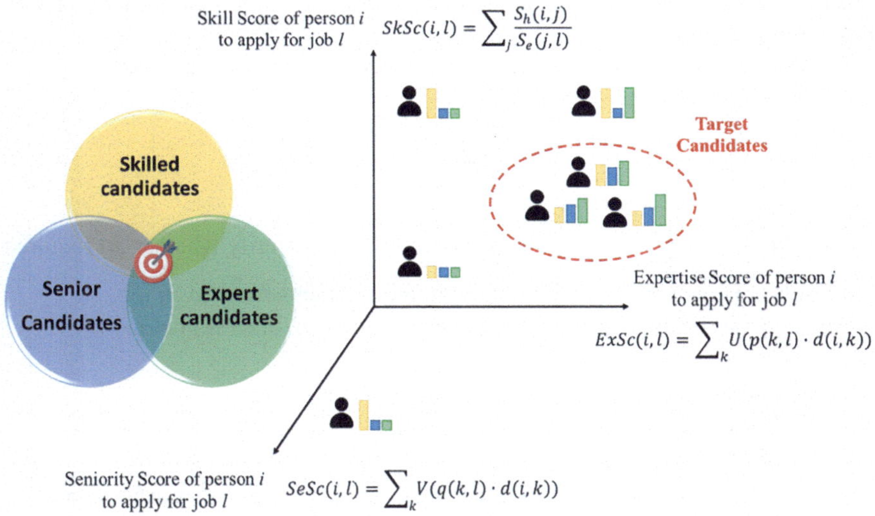

Fig. 8.4 Utilization of matching to identify candidates

adapted to the needs of any specific context, in one example, each score below is comprised in the range [0,10]; the score 0 corresponds to a total lack of skill, while the score 10 indicates the skill level of a highly trained and experienced professional. Thus, we define the following variables:

- $S_h(i, j)$ the level of skill j held by a person i
- $S_e(j, l)$ the expected level skill of skill j to obtain a job l

Below, we propose several approaches for measuring the combined skill of a person derived from individual skills. In a first proposal, the skill score SkSc of person i on job l is defined as

$$SkSc(i, l) = \sum_j \frac{S_h(i, j)}{S_e(j, l)} \quad (1a)$$

While approach (1a) may be suitable for an average candidate with an average score for each individual skill, it may create a misleading total score for extreme cases where certain skills are very high while other skills are very low. This is taken into account in the following refined approach, where an upper bound for normalized skills are introduced:

$$SkSc(i, l) = \sum_j UB\left[\frac{S_h(i, j)}{S_e(j, l)}\right] \quad (1b)$$

With $UB[x] = \begin{cases} x \ if \ x < x0 \\ x0 \ if \ x \geq x0 \end{cases}$. Hereby, "$x0$" indicates an upper bound for the normalized skill, for example "$x0 = 1.5$". As a result, a person being extremely skilled in one domain—more than required for the job—cannot hide a total lack of skill in other domains.

In a third alternative, a binary decision is taken on the normalized skill:

$$SkSc(i, l) = \sum_j BIN\left[\frac{S_h(i, j)}{S_e(j, l)}\right] \quad (1c)$$

With $BIN[x] = \begin{cases} 0 \ if \ x < 1 \\ 1 \ if \ x \geq 1 \end{cases}$. This metric helps to identify whether all required skill levels are equal or above the minimum requirements for a given job.

We compare the three proposed approaches (1a), (1b) and (1c) below:

Approach	Advantages	Disadvantages
(1a)	Maintains outstanding individual skills in the final score	One outstanding skill may obfuscate a lack of other essential skills
(1b)	Maintains outstanding individual skills in final score to a certain extend	Outstanding skills may obfuscate a lack of other essential skills to a certain extent
(1c)	Final score indicates whether all required skills are above the required skill level	The final skill score does not value extremely high individual skills

Then, we identify among the career of the considered person whether it has a concrete experience in the given domain and how long it is. We also define customizable limits to identify candidates with the proper experience (it may be inadequate to select someone with too low or too high experience). In this perspective, we define the following variables:

- $p(k, l)$ the proximity of domain between a job of experience k exerced by the person and the target job l (it corresponds to a percentage of adequacy between jobs)
- $d(i, k)$ the duration of the experience of job k by person i

- $v(i, k, l) = p(k, l) \cdot d(i, k)$ the value of experience k for a person i to obtain a job l in the domain perspective
- $U(v(.))$ the utility function such as:
  - $U(v(d)) = 0$ when d < Min
  - $U(v(d)) =$ v(d) when Min < d < Max
  - $U(v(d)) =$ v(Max) when d < Min [Markus: shouldn't that be d > Max?]

The expertise score ExSc of person i on job l may be defined as:

$$ExSc(i, l) = \sum_k U(p(k, l) \cdot d(i, k)) \tag{8.2}$$

Equation (8.2) calculates the sum of all experiences (in related domains projected onto the target job domain) multiplied by the duration of the individual experience.

The Utility function *U(v(d))* ensures that only relevant experiences are considered. Afterwards, we propose to introduce a score that takes into account the seniority of the candidate. This aims to promote a candidate that already exerced a demanding job even if it was not directly in the same domain. For example, a former member of a cabinet office is more likely to lead a large organization, since it faced a complex situation with political implications. To achievthis objective, we define the following variables:

- $q(k, l)$ the proximity of seniority between a job of experience k exerced by the person and the target job l (it corresponds to a percentage of adequacy between jobs)
- $d(i, k)$ the duration of the experience of job k by person i
- $w(i, k, l) = p(k, l) \cdot d(i, k)$ the value of experience k for a person i to obtain a job l in the seniority perspective
- $V(w(.))$ the utility function such as:
  - $V(w(d)) = 0$ when d < Min
  - $V(w(d)) =$ v(d) when Min < d < Max
  - $V(w(d)) =$ v(Max) when d > = Max

The seniority score SeSc of person i on job l may be defined as

$$SeSc(i, l) = \sum_k V(q(k, l) \cdot d(i, k)) \tag{8.3}$$

Equation (8.3) calculates the sum of all experiences (in related domains projected onto the target job domain) multiplied by the duration of the individual experience. The Utility function *V(w(d))* ensures that only relevant experiences are considered. Finally, we combine the three separate scores by introducing the possibility of introducing a priority through a triplet (α, β, γ) such as $\alpha + \beta + \gamma = 3$. The interest in preserving a fixed number is to facilitate the comparison between different options. However, another configuration is possible. A first simulation should set the three parameters equal to 1. The final equation is:

$$S(i, l) = \alpha SkSc(i, l) + \beta ExSc(i, l) + \gamma SeSc(i, l) \tag{8.4}$$

The weighted sum thus guarantees that a balance between the various scores is being found while the choice of (α, β, γ) allows for a job specific fine-tuning of the priorities for a specific position.

8.4 End-to-End Trustworthiness

The correct operation of an automated Government and inherent related processes is key to its success and viability. Figure 8.3 illustrates how a variety of stakeholders are interacting with the system, either as a source of information, as part of a processing step or as a final user of generated outputs. Such an approach is open to a variety of security vulnerabilities, including, for example, a so-called "Man-in-the-Middle" attack where a malicious stakeholder, intercepts some intermediate communication and alters the content as illustrated in Fig. 8.5.

Further potential vulnerabilities include the provision of erroneous input to confuse the system and possibly trigger malfunction, the inclusion of malware et many more.

In the context of an automated Government vision, such attacks are especially harmful since they may undermine the operation and thus the legitimity of the Government. Great care thus needs to be taken to implement appropriate safeguards and verification processes.

In the context of the present paper, the authors suggest the following requirements:

(1) At each point in the chain, any previous steps need to be fully tracable; i.e., we need to be able to identify the originator of any data being used for processing and all processing steps need to be accessible and verifiable;

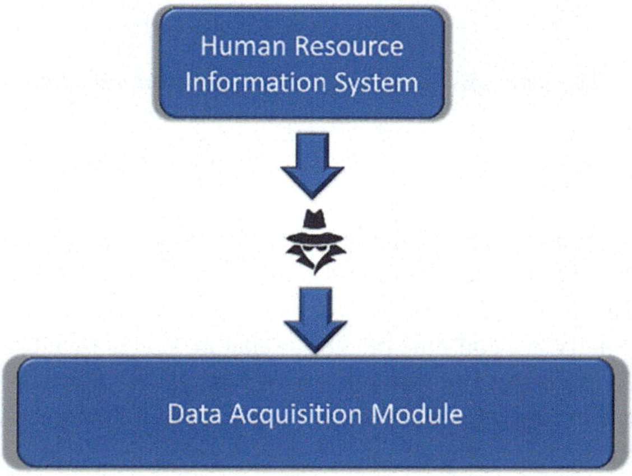

Fig. 8.5 Illustration of a "Man-in-the-Middle" attack

(2) Data processing must rely on secure computation principles, ideally all data is encrypted at all times—except when being processed; processing is done in a trusted execution environment which receives encrypted data as input and provides encrypted data as output (i.e., decrypted data is only accessible within a highly secured and trusted component);
(3) Data being transferred between stakeholders needs to be highly protected, since it is potentially exposed to a variety of attacks.

Item 1) is proposed to be achieved through a blockchain approach; i.e., documentation on any additional processing step is added to an available blockchain such that the final consumer of processed data has full knowledge of any intermediate step and is able to interact with related stakeholders (executing the intermediate processing steps). In case of doubt, the processing results can be cross-checked.

Item 2) is achieved by relating on latest State-of-the-Art secure computation and confidential computing solutions such as [27].

Item 3) is proposed to be addressed through an appropriate Certificate Management framework. A cryptographically protected certificate is created at the output of each intermediate processing step and is registered to a trusted Certificate Management Database/Service Provider. Prior to the next processing step, the Certificate is verified and its validity is being checked by reaching out to the trusted Certificate Management Database/Service Provider. Only if the Certificate is validated, the next processing step is undertaken. The basic principle is illustrated in Fig. 8.6 for a selected part of the chain:

Concerning the End-to-End security and trustworthiness of the system, we need to consider entirely new attack strategies for AI/ML based learning and related decision making. For example, an attacker may temper the learning data that is acquired at

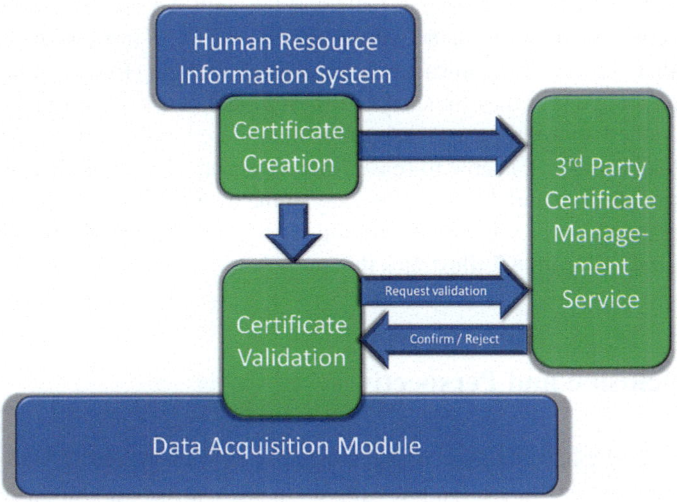

Fig. 8.6 Principle of a certificate framework

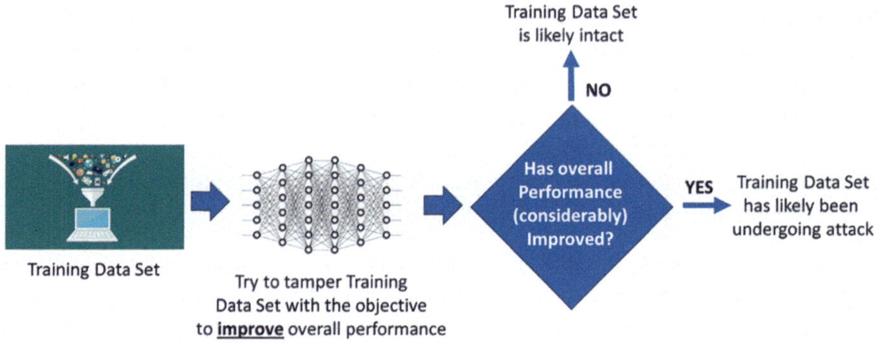

Fig. 8.7 Detection of a likely attack on the original training data set

the beginning of the learning process and typcially continued over the lifetime of the AI/ML model in order to refine and adapt the model to an evolving environment.

While such an attacker may target a reduction of the overall AI/ML performance through tampering of the learning data, it is often challenging to identify which learning data sets have been compromised and which are not. This is becoming even more difficult in the context of extremely large data sets.

One strategy for detection of a potential attack on the training data set relies on the usage of techniques similar to the ones applied by an attacker. Indeed, we take the available training data set and have to objective to identify whether it has been undergoing an attack with the objective to lower its relevance and thus to negatively impact the overall efficiency of the AI system. We then apply a (deep neural network based or similar) "attack" with the objective of improving the quality of the Training Data Set and thus to improve the performance of the final AI system; to some extent, we do exactly the same as a potential attacker but with the opposite objective, i.e. instead of reducing the performance of the system, we try to improve it.

In case that our tentative is unsuccesful, i.e. it is challenging to improve the overall training data set and thus the efficiency of the AI system, we assume that the training data set is indeed appropriate and has not been tampered with. In the opposite case, i.e. our "attack" strategy is able to (easily) improve the overall validity and quality of the training data set and thus the efficiency of the overall AI system, then we must assume that there is an issue with the training data set and it has likely undergone a malicious attack. Figure 8.7 illustrates this proposal.

8.5 Conclusion and Perspectives

In this chapter, we have detailed and evaluated several measures to modernize the processes for Human Resources through Artificial Intelligence. First, the authors provide a comprehensive literature review concerning the potential usage of AI in the specific context of e-Government and the processes for Human Resources. Then, the authors propose a new framework to improve the association between people and jobs. This aims to optimize the recruitment of candidates as well as career management. Finally, the authors describe a process to ensure end-to-end trustworthiness and enhance the security of e-Government.

This chapter aims to propose new techniques and approaches to develop the usage of Artificial Intelligence top optimize HR for government services. While the proposals are promising, they need to be further developed and implemented in concrete situations to provide operational feedbacks and facilitate further implementation. The authors will propose to young researchers or students to define a data set and initiate an experiment. Then, it will be possible to experiment it on a real government administration population.

References

1. Danziger, J. N., Kraemer, K. L., & King, J. L. (1978). An assessment of computer technology in U.S. local governments. *Urban Systems, 3*(1), 21–37. ISSN 0147-8001. https://doi.org/10.1016/0147-8001(78)90004-9
2. Kraemer, K. L., et al. (1978). *Local government and information technology in the United States*. Paris: OECD Informatics Studies #12.
3. Yildiz, M. (2007). e-Government research: Reviewing the literature, limitations, and ways forward, *Government Information Quarterly, 24*(3), 646–665. ISSN 0740-624X. https://doi.org/10.1016/j.giq.2007.01.002
4. Grönlund, Å. (2010). Ten years of e-Government: The 'End of History' and new beginning. In M. A. Wimmer, J. L. Chappelet, M. Janssen, & H. J. Scholl (Eds.), *Electronic government*. EGOV 2010. Lecture notes in computer science, vol. 6228. Springer. https://doi.org/10.1007/978-3-642-14799-9_2
5. Bhuvana, M., Ramkumar, A., & Neeraja, B. (2023). Machine learning technique for predicting the rural citizens' trust on using e-Governance health care applications during COVID-19. In C. Gaie & M. Mehta (Eds.), *Recent advances in data and algorithms for e-Government*. Artificial intelligence-enhanced software and systems engineering, vol. 5. Springer. https://doi.org/10.1007/978-3-031-22408-9_6
6. Gaie, C., & Mueck, !M. (2023). Enhancing government actions against Covid-19 using computer science. In C. Gaie & M. Mehta (Eds.), *Recent advances in data and algorithms for e-Government*. Artificial intelligence-enhanced software and systems engineering, vol. 5. Springer. https://doi.org/10.1007/978-3-031-22408-9_8
7. Gkikas, D. C., Theodoridis, P. K., & Gkikas, M. C. (2023). Artificial Intelligence (AI) use for e-Governance in agriculture: Exploring the bioeconomy landscape. In C. Gaie & M. Mehta (Eds.), *Recent advances in data and algorithms for e-Government*. Artificial intelligence-enhanced software and systems engineering, vol. 5. Springer. https://doi.org/10.1007/978-3-031-22408-9_7
8. Gaie, C. (2023). Struggling against tax fraud, a holistic approach using artificial intelligence. In C. Gaie & M. Mehta (Eds.), *Recent advances in data and algorithms for e-Government*. Artificial intelligence-enhanced software and systems engineering, vol 5. Springer. https://doi.org/10.1007/978-3-031-22408-9_4
9. Neves, F. R., & Silva, P. B. (2023). From paper to digital: e-Government's evolution and pitfalls in Brazil. In C. Gaie & M. Mehta (Eds.), *Recent advances in data and algorithms for e-Government*. Artificial intelligence-enhanced software and systems engineering, vol. 5. Springer. https://doi.org/10.1007/978-3-031-22408-9_9

10. Wirtz, B. W., Weyerer, J. C., & Geyer, C. (2019). Artificial intelligence and the public sector—Applications and challenges. *International Journal of Public Administration, 42*(7), 596–615. https://doi.org/10.1080/01900692.2018.1498103
11. Al-Mushayt, O. S. (2019). Automating e-Government services with artificial intelligence. *IEEE Access, 7*, 146821–146829. https://doi.org/10.1109/ACCESS.2019.2946204
12. Kaya, T. (2019). Artificial intelligence driven e-Government: The engage model to improve e-Decision making. https://doi.org/10.34190/ECDG.19.054
13. Al-Besher, A., & Kumar, K. (2022). Use of artificial intelligence to enhance e-Government services. *Measurement: Sensors, 24*, 100484. ISSN 2665-9174. https://doi.org/10.1016/j.measen.2022.100484
14. Thomas, J. C. (2003). Gregory Streib, The new face of government: Citizen-initiated contacts in the era of e-Government. *Journal of Public Administration Research and Theory, 13*(1), 83–102. https://doi.org/10.1093/jopart/mug010
15. Sanmukhiya, C. (2019). Predicting e-Government use in Mauritius: Non-parametric procedures. In *International Journal of Recent Technology and Engineering. Blue Eyes Intelligence Engineering and Sciences Engineering and Sciences Publication—BEIESP*. https://doi.org/10.35940/ijrte.b1114.0982s919
16. Anthopoulos, L. G., Siozos, P., & Tsoukalas, I. A. (2007). Applying participatory design and collaboration in digital public services for discovering and re-designing e-Government services. *Government Information Quarterly, 24*(2), 353–376. ISSN 0740-624X. https://doi.org/10.1016/j.giq.2006.07.018
17. Al-hassan, M., Lu, H., & Lu, J. (2011). Personalized e-Government services: Tourism recommender system framework. In J. Filipe & J. Cordeiro (Eds.), *Web information systems and technologies*. WEBIST 2010. Lecture notes in business information processing, vol. 75. Springer. https://doi.org/10.1007/978-3-642-22810-0_13
18. Mostafa, M. M., & El-Masry, A. A. (2013). Citizens as consumers: Profiling e-Government services' users in Egypt via data mining techniques. *International Journal of Information Management, 33*(4), 627–641. ISSN 0268-4012. https://doi.org/10.1016/j.ijinfomgt.2013.03.007
19. Maksimova, M., Solvak, M., & Krimmer, R. (2021). Data-driven personalized e-Government services: Literature review and case study. In: N. Edelmann, et al. *Electronic participation*. ePart 2021. Lecture notes in computer science, vol. 12849. Springer. https://doi.org/10.1007/978-3-030-82824-0_12
20. McCarthy, J. (1960, April). Recursive functions of symbolic expressions and their computation by machine, Part I. *Communications of ACM, 3*(4), 184–195. https://doi.org/10.1145/367177.367199
21. Votto, A. M., Valecha, R., Najafirad, P., & Raghav Rao, H. (2021). Artificial intelligence in tactical human resource management: A systematic literature review. *International Journal of Information Management Data Insights, 1*(2), 100047. ISSN 2667-0968. https://doi.org/10.1016/j.jjimei.2021.100047
22. Vrontis, D., Christofi, M., Pereira, V., Tarba, S., Makrides, A., & Trichina, E. (2022). Artificial intelligence, robotics, advanced technologies and human resource management: A systematic review. *The International Journal of Human Resource Management, 33*(6), 1237–1266. https://doi.org/10.1080/09585192.2020.1871398
23. Pan, Y., Froese, F., Liu, N., Hu, Y., & Ye, M. (2022). The adoption of Artificial Intelligence in employee recruitment: The influence of contextual factors. *The International Journal of Human Resource Management, 33*(6), 1125–1147. https://doi.org/10.1080/09585192.2021.1879206
24. Baker, J. (2012). The technology–organization–environment framework. In Y. Dwivedi, M. Wade & S. Schneberger (Eds.), *Information Systems Theory. Integrated Series in Information Systems*, vol. 28. Springer. https://doi.org/10.1007/978-1-4419-6108-2_12
25. Gong, Y., Zhao, M., Wang, Q., & Lv, Z. (2022). Design and interactive performance of human resource management system based on artificial intelligence. *PLoS ONE, 17*(1), e0262398. https://doi.org/10.1371/journal.pone.0262398

26. Huang, X., Yang, F., Zheng, J., Feng, C., & Zhang, L. (2023). Personalized human resource management via HR analytics and artificial intelligence: Theory and implications. *Asia Pacific Management Review*. ISSN 1029-3132. https://doi.org/10.1016/j.apmrv.2023.04.004
27. Confidential Computing. https://www.intel.com/content/www/us/en/security/confidential-computing.html
28. Jordan, E., Shin, D. E., Leekha, S., & Azarm, S. (2021). Optimization in the context of COVID-19 prediction and control: A literature review. *IEEE Access, 9*, 130072–130093. https://doi.org/10.1109/ACCESS.2021.3113812
29. Grönlund, Å., & Horan, T. (2005). Introducing e-Gov: History, definitions, and issues. *Communications of the Association for Information Systems, 15*. https://doi.org/10.17705/1CAIS.01539
30. Veljković, N., Bogdanović-Dinić, S., & Stoimenov, L. (2012). Building e-Government 2.0—A step forward in bringing government closer to citizens. *Journal of e-Government Studies and Best Practices, 2012*, 18. Article ID 770164. https://doi.org/10.5171/2012.770164
31. Mettler, T. The road to digital and smart government in Switzerland. In A. Ladner, N. Soguel, Y. Emery, S. Weerts, & S. Nahrath (Eds.), *Swiss public administration. Governance and public management*. Palgrave Macmillan. https://doi.org/10.1007/978-3-319-92381-9_10
32. Mossey, S., & Manoharan, A. (2018). e-Government policy and legislation: A chronological timeline. In Farazmand, A. (Eds.), *Global encyclopedia of public administration, public policy, and governance*. Springer. https://doi.org/10.1007/978-3-319-31816-5_3303-1
33. Venkatesh, V., Morris, M. G., Davis, G. B., & Davis, F. D. (2003). User acceptance of information technology: Toward a unified view. *MIS Quarterly, 27*(3), 425–478. https://doi.org/10.2307/30036540

Chapter 9
Fuzzy Logic Architecture for Availing E-Governance Healthcare Services by Rural Citizens

M. Bhuvana and A. Ramkumar

Abstract The Larger part of the Indian populace dwells in rustic towns in our society. It is detailed from the World Bank that 66 percent of the complete populace in India is secured by provincial individuals. Subsequently any administrations encouraged by the Indian government ought to reach the provincial individuals straightforwardly. E-governance is the application planned and created by the government to offer its administrations and fulfill the essential prerequisites of the rustic individuals through Data and Communication Innovation (ICT) across the globe. Developed countries such as the United States, China and the United Kingdom have implemented effective e-Government services. Indeed, in spite of the fact that the activities are taken by the government to offer its best administrations to reach the rustic individuals, the recurrence of evaluating and utilizing it beneficially is still missing in rustic towns. This think about has inspected on recognizing the actualities that are emphatically related with the utilization measurements of provincial individuals towards e-governance administrations. Based on the measurements from the writing study the analysts have planned the Fuzzy framework utilizing R Programming Dialect to analyze and anticipate the real utilization of e-governance administrations. Nine Fuzzy rules have been organized and tried with four distinctive sets of illustrations. The comes about produced by the Fuzzy framework makes a difference the government authorities and the policymakers for distinguishing the outlets of the rustic individuals for beneficially getting to e-governance administrations.

Keywords Actual usage · E-governance · Fuzzy system · ICT · Rural citizens

M. Bhuvana (✉) · A. Ramkumar
SRM Institute of Science and Technology (SRMIST), Kattankulathur, India
e-mail: bhuvi.tejal@gmail.com

A. Ramkumar
e-mail: ramkumaa@srmist.edu.in

C. Gaie and M. Mehta (eds.), *Transforming Public Services—Combining Data and Algorithms to Fulfil Citizen's Expectations*, Intelligent Systems Reference Library 252, https://doi.org/10.1007/978-3-031-55575-6_9

9.1 Introduction

E-Governance is stated as the method of interaction with the government services electronically. The transition of services from the government through Information and Communication Technology is found to be a drastic increase [1].

It is narrated the fuzzy system has four dissimilar areas like Business Process Management, Information and Communication Technology, and e-governance. Furthermore, it is the distribution channel between state, and national government services through ICT [2, 3]. It is the determination of facilitating the information and the public services to the rural citizens through ICT [4]. It provides not only the public services to the people and also create a bond between the businesses, government, government agencies and the workers [3, 5].

9.1.1 Global Concern of E-Governance Services

Developed countries such as the United States, China, and the United Kingdom have implemented effective e-Government services [6]. Table 9.1 describes the global E-Participation Index survey report in 2022. It shows that Japan, Australia, and Estonia have gained the top three positions in access to e-governance services. However, usage and awareness of e-Government services in India is still evolving. 69.8% of India's population lives in rural and remote villages [7, 8]. The researchers have described that the Indian Government has made several projects regarding the development of e-governance services out of which e-Kranti has the major component that has launched recently under Digital India Programme that has made a special focus on multi state and national Agri-based projects under e-governance [9, 10].

Table 9.1 2022 E-participation index

EPI rank in 2022	Country	EPI value in 2022	EPI rank in 2020	Change n EPI rank from 2020 to 2022
1	Japan	1.0000	4	+7
2	Australia	0.9886	9	+7
3	Estonia	0.9773	1	−2
3	Singapore	0.9773	6	+3
5	Netherlands	0.9659	9	+4
6	Finland	0.9545	14	+8
6	New Zealand	0.9545	4	−2
6	United Kingdom of Great Britain and Northern reland	0.9545	6	0

9.1.2 E-Governance in Indian Context

The Government of India has made great efforts to increase awareness of e-Government services, such as the establishment of the Digital India project [11, 12]. Various research Studies have shown that the behaviour of the citizens residing in rural areas and plays a vital part in retrieving government services electronically [13]. Several Researchers used the Technology Acceptance Model (TAM) to develop a conceptual framework for measuring the actual use of banking services [14, 15]. Rural people's financial knowledge pays the price for making the best decisions related to financial planning [16]. Literacy on ICT makes opportunities for citizens from rural background to learn about government services online [17]. e-Government services make all kinds of business and knowledge and experience sharing transparent and simple for people. Thus, ICTs have successfully contributed to rural development by facilitating door-to-door access to rural communities [9, 18]. The authors have predicted the exact dimension relating to level of trust among the citizens from rural areas for accessing the various applications of e-governance services. The researchers have constructed the theoretical framework for accessing the accurate dimensions towards rural citizens trust for using the several e-governance applications [19]. Authors have made a theoretical study on e-Governance through a muti-dimensional construct that determines hyper integrated network, empowered citizenship and evolutionary system architecture as three dimensions for utilizing e-Government services for future requirements [2]. Several authors have discussed about the various components of e-governance services such as "Interactivity", "Building Blocks", and "Interoperability", are described as the major components of e-governance services [20]. The research study have considered classical approach to determine the diverse ranges of locations where the e-governance volume has been accompanied. "Digitalization, Legal, User's acceptability, Literacy Level and Awareness" are the five major sub areas focused by the researchers to understand the existence of e-governance services across various areas [21].

The authors have built the E-govDR Model (e-Government Design Research Model) to review the designs of e-governance services. The researchers have identified the various task and activities involved in e-governance framework that supports the decision support tasks [22]. The researchers have constructed the WSF (Works System Framework) for understanding the functioning of e-Government work system (e-GovWSF). The authors have discussed about 12 different elements that elaborated the description of association between internal, external, semi-external work systems of e-Government [23]. The research paper has targeted to examine the various determinants of e-governance between the decades 2000–2015. It is found that financial, technological, security, privacy, and human resources are the major and exact dimensions of e-governance system [24].

The researchers have examined the development, creation and implementation of municipal institutions during COVID-19 Pandemic in Russian health care system. Case Study method has been adopted in the study. The study has elaborated the findings of the government platform that has been constructed throughout the pandemic of COVID-19 and technologies usage to involve the rural poor [25, 26]. Fuzzy logic is developed and verified with 6 different pairs of fuzzy rules to realize the system in present research study. The variables given as inputs are mapped to values and attributes that describe the act of the system. The outcome of the system help officials of the government and policy makers determine access to government e-participation for the rural citizen population.

9.2 Review of Literature

9.2.1 Evidence Related to Information and Communication Technology

Several applications have been developed and designed to ensure citizens residing in rural villages for enabling e-governance system. Several applications have been built for the rural citizens that can be easily accessible to satisfy their needs in a transparent manner [27, 28]. The word "awareness" means a person who understands the activities of others who are responsible for their activities. The authors criticize the term "consciousness" as "unpleasant or vague" and believe that consciousness "can hardly be a concept lacking in quality" [29]. In this study, the authors emphasized that human support or self-awareness ultimately improves team or individual performance [30]. The researchers explained that ICT literacy is very low, especially among people living in rural areas. Lack of understanding about e-governance services, association between administration and rural population is found to be challenges for delivering the services [31]. The study suggests that programs related to capacity building should be developed by various financial institutions to promote the use of ICT in rural areas. It is important to discuss the use of ICT for rural development and raising the awareness of rural people [32].

9.2.2 Literacy

ICT literacy is defined as the integration, management, evaluation and creation of digital technologies, communication tools, and access to websites and data for the productive functioning of the society [19]. Literacy on ICT can be determined describing a person's ability to use technology and communicate effectively [33]. The literacy rate in rural India is 68.9%. Therefore, it is very important for illiterate people to find the necessary technological knowledge to use e-Government services [34]. This technological literacy can be increased by the authorities working in government institutions by creating or establishing digital projects [35]. Capacity building programs are the best way to train rural people to use e-Government services. In e-Government services, the public sector leverages the availability of information and communication technology (ICT) to bring real change to the country [36].

9.2.3 *Health Care Services Usage Under E-Governance*

The word "actual use" is defined as a person's behavioural response as measured by a person's real-life behaviour. It is defined by the sum of time spent and the occurrence of operations [37]. This study examines the "emotional behaviour" dimension of online shopping by consumers in Turkey. The authors integrated the Technology Acceptance Model (TAM) and Planned Behaviour Theory (TPB) to enhance their research work and also identified processes associated with actual use in internet shopping [38]. The authors measure consumer behaviour when using online financial tools. The preliminary data for this current study were composed by superior analysts in the business field in business school. Researchers have exposed a positive association between the variable's "attitude" and "actual use" of banking through online [39]. The researchers confirmed that the "directly beneficial change" difference affects people's attitudes and behaviours towards the actual use of the banking services producer [40]. The use of e-Government services is being investigated by using public interest through intermediation. It is said that public relations play a vital role in evaluating the services usage under e-governance. Researcher have utilized approval, approval and quality services as determinants to quantify satisfaction of the public towards e-Government services [41].

9.3 Methods

The current study is grounded on the analysis of statistics from the secondary source composed from previous studies from previous studies. This study identifies factors and factors that predict the actual use of government e-Government health services by people living in underserved areas of our society. By analyzing previous data, the researchers linked two factors, 'ICT knowledge' and 'knowledge' level, to determine the use of e-Government healthcare services. This study uses "prediction science" to develop rules in fuzzy logic algorithm. This estimation method was introduced in 90s to stress on the uncertainty related with the components used in the conceptual model and duplicability [23, 42]. Furthermore, not only are production uncertainties important, but also uncertainties that are important to the project, organization, or environment. Thus, according to research data, researchers have found that Internet penetration in the year 2020 in India has increased from 48.4% to 50% [22]. The situation is estimated that the internet users are found to be 290 million in rural India [43]. However, the rural development total score for national e-Government health service evaluation is 0.56 and 0.37 for health and family health [21]. Despite the rise in the amount of internet users in rural areas, it is difficult for the rural population's knowledge and literacy level to get the lowest score in the general evaluation of e-Government services.

9.4 Architecture of Fuzzy Logic

The fundamental framework or architecture of fuzzy logic systems consists of four different components as shown in Fig. 9.1 [33]. The first component "Rule Base" includes the descriptions and "if then" situations that involves the experts to support in decision making. Present inventions in the fuzzy logic theories facilitate various productive methods for optimizing and designing fuzzy controllers. These innovations help in the advances and improvements that helps in reducing the uncertain rule sets.

The second component "Fuzzification" determines the process that converts the numbers into the fuzzy sets. In this component we could measure the exact inputs and diffuse them to the control system for further processing [33]. The component "Fuzzification" splits the input signals into five different states as shown in Table 9.2.

The third component "Inference Engine" processes the data, that is also known as a primary component of "FLS (Fuzzy Logic System)". This component measures the identical level that exists between present fuzzy logic input and all the rules and selects the rules that has to be executed based on the present input field. After the execution of all the rules, control actions will be created and grouped. The fourth component "De-fuzzification" transforms the generated fuzzy sets by the inference engine and converted into a crisp value. Several de-fuzzification techniques has been

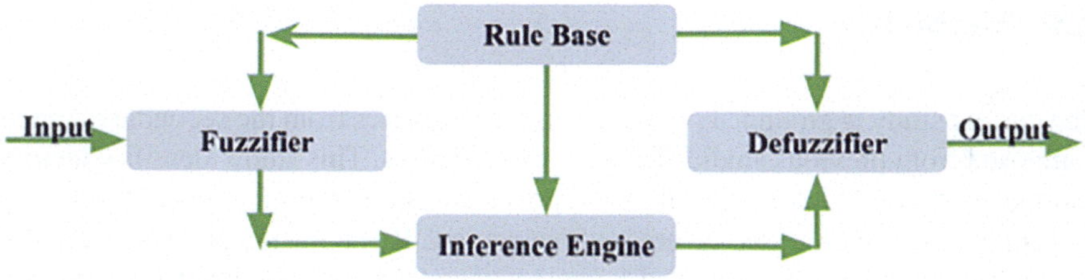

Fig. 9.1 Fuzzy logic architecture

Table 9.2 Fuzzification states

LP	X is large positive
MP	X is medium positive
S	Small
MN	X is medium negative
LN	X is large negative

used to minimize the error and it is said to be the final stage for developing the fuzzy logic system [33].

The "MFs (Membership Functions)" are the important and essential components of the Fuzzy Set Theory that has been implemented using fuzzy logic controller. It is represented as the graph that summarizes the values of membership between 0 and 1 in the given input space for every point. The main objective of the membership function is to plot the fuzzy linguistic terms with the non-fuzzy input values. It also enables us to picture the fuzzy set and count the linguistic words [33].

In this study, algorithm for fuzzy logic was developed by utilizing the programming language called RStudio. The primary goal of this project to develop a system of communication in R to regulate the occurrence of healthcare services under e-governance in India especially the rural villages. It computes and assesses the rules of fuzzy sets for decision making using various input tests. However, it does not directly use the data model to describe fuzzy problems [20]. This research report uses "literacy" as an indicator and to provide context for the "actual use" of e-Government health services in rural areas. In order to create a fuzzy system, nine fuzzy rules were determined based on the assumption of not using/moderately/frequently using e-Government services.

Advantages and Limitations of Fuzzy Logic System

The advantages of fuzzy logic systems are:

- The fuzzy logic system could handle aby kinds of incorrected, disorganized, and cluttered data.
- It is quite simple to understand and built.
- It is completely based on theoretical mathematical principles.
- It facilitates highly productive solutions that complicates several problems for the human beings to take up the right decisions.
- Limited information and storage space is required for briefing the algorithms [33].

The disadvantages of fuzzy logic systems are:

- It has a slow runtime and it generates the output quite slowly.
- In Artificial Intelligence, Fuzzy Logic may not be appropriate for certain situations for demanding high accuracy.
- A fuzzy based system needs appropriate equipment testing for validation and conformation.
- Fuzzy Logic usually utilizes imprecise and accurate data [33].

9.4.1 Fuzzy System for the Present Study

The authors of the present study have used methods to use fuzzy systems in various areas of application development. Blurring, inference or rules, and clarification are considered the 3 main manifestations for the methods adopted for fuzzy logic. Blurring remains taking attributes and turning it into variables of fuzzy logic. The subsequent step is to create rules through thinking about functions of the system, and the de-fuzzification is the former stage, which changes the result of the system for a more precise change. Fuzzy logic has been successfully used in many fields such as medicine, business, agriculture, politics, operations science and environmental science [44]. The term "fuzzy logic" has been used and expanded in many studies to better understand the concept of the body [45]. The results are further expanded by considering the behaviour according to the behaviour in the study. Researchers scoured the data of applications developed by neuro-fuzzy machines. Scientists use fuzzy logic techniques from agriculture for pest control. Specialized methods have been developed to analyse soil-based harvests [46]. The research study has, four input and five output variables and 8 fuzzy rules were obtained for the selected variables in this study. The type of fabric, soil and fabric quality were considered as variables by experts for production [47].

9.4.2 Determination of Variables and Attributes

'ICT awareness' and 'information awareness' variables were used to analyze the use case of e-Government services. The system of fuzzy logic has 9 rules with 3 variables as shown in Table 9.3.

Each variable in the system has 3 different attributes. The variable "info" has the values "low", "medium", "high" with values 90, 70 & 30. The variable "ICT Awareness" have the characteristics "Bad", "Moderate" and "Excellent" as its values 30, 60 and 80. The last variable "Use" is "not used", "Moderate" and "Frequent" according to the characteristic values of the state of the system. "Use" means 90, 30 and 60.

Table 9.3 Variable names and its attributes

S. no.	Variable name	Attributes
1	Usage	1. no.Usage 2. Moderate.Usage 3. Frequent.usage
2	ICT Awareness	1. Poor 2. Average 3. Excellent
3	Literacy	1. Low 2. Medium 3. High

9.4.3 Formation of Fuzzy Rules

The subsequent step is to determine the rules of the fuzzy system. The rules of the fuzzy system lead to the integration of variables. Nine fuzzy rules listed below have been developed for this system. Nine fuzzy rules are defined for fuzzy systems:

S. no.	Condition for literacy level	Condition for ICT awareness	Outcome of E-governance usage
1	Low	Poor	No usage
2	Low	Average	Moderate usage
3	Low	Excellent	Frequent usage
4	High	Poor	No usage
5	High	Average	Moderate usage
6	High	Excellent	Frequent usage
7	Medium	Poor	No usage
8	Medium	Average	Moderate usage
9	Medium	Excellent	Frequent usage

9.4.4 Formation of Fuzzy System

Figure 9.2 shows the plan developed for rural population to use e-Government services. By doing fuzzy_system() and set() the plans appear to overlap towards different "health information". These demonstrations overlap between Medium and High. Therefore, if the health awareness is 75, the user can be considered as moderate and heavy.

9.4.5 Testing the System with Fuzzy Rules

The four different inputs are given to test the fuzzy system designed for the study and the illustrations are recorded below.

Example 1
#When Literacy on Health is 75, and Awareness on ICT is 25.

Figure 9.3 signifies the yield of the system. Grounded on the above criteria, if information of health is moderate that is 75 and the knowledge of ICT is poor that is 25, then the utilization rate is 30 from the e-governance services, describing the "not used" quality. It was decided that if the health information is recognized and the level of knowledge is good, entering the government e-Government will not be beneficial.

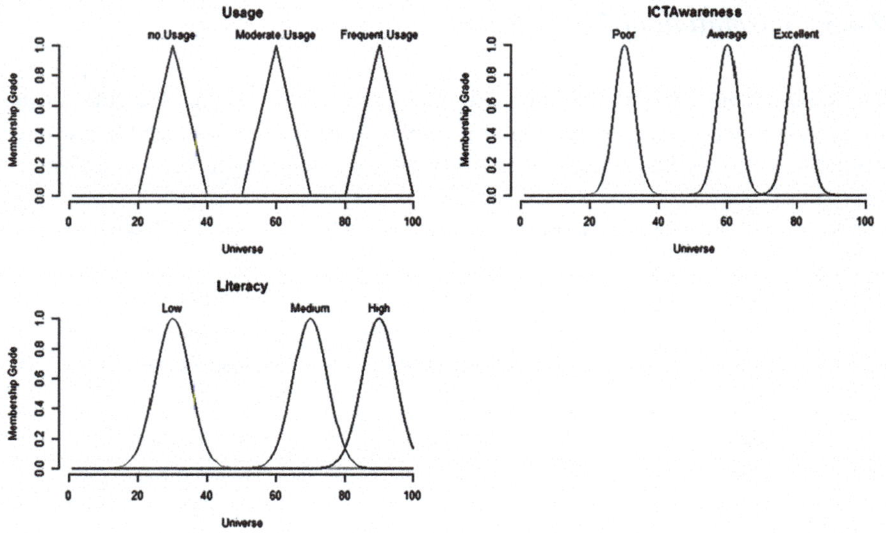

Fig. 9.2 Fuzzy system for health care usage

Therefore, the system for fuzzy rules is verified with various contributions from the instances below.

Example 2
 # When the Literacy on Health is 70 and Awareness on ICT is 60.

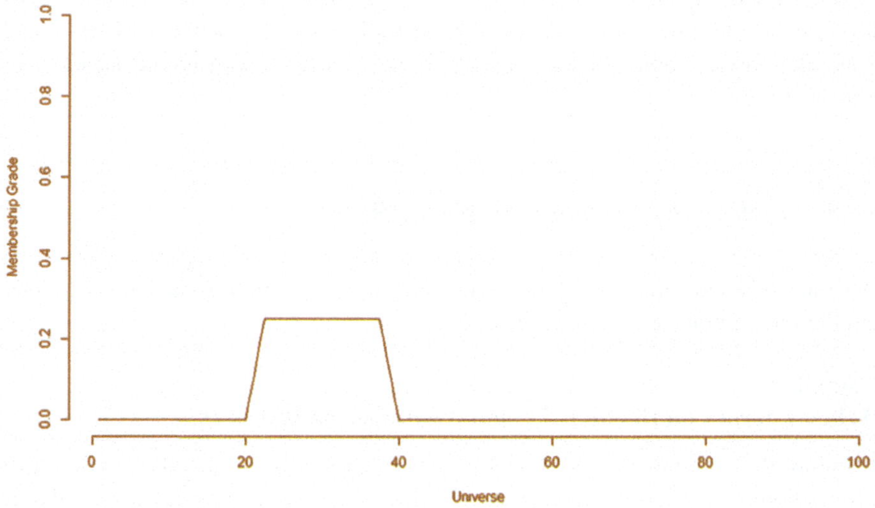

Fig. 9.3 Health literacy is 75, and awareness on ICT is 25

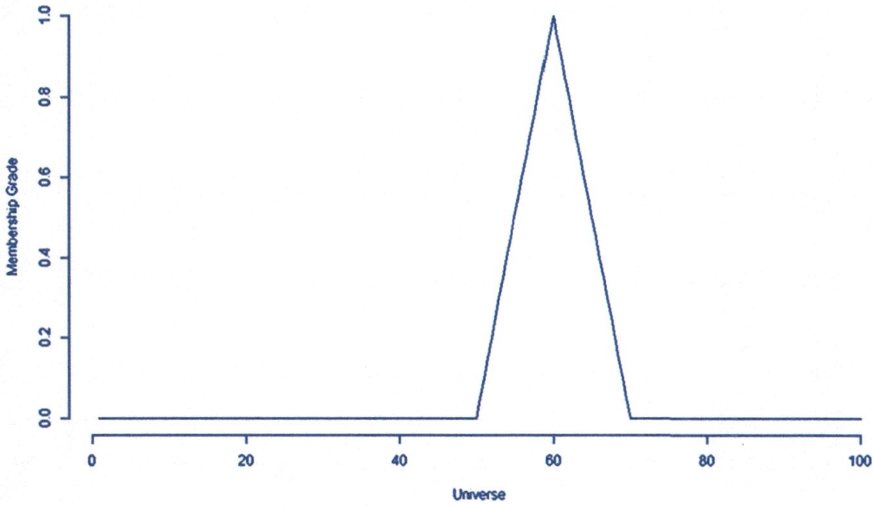

Fig. 9.4 Literacy on health is 70 and awareness on ICT is 60

Figure 9.4 display the outcome of the system for the rules of fuzzy system. Grounded on the system with the value of Average "70" as Literacy Level and Average "60" as Knowledge on ICT, the amount of practice of services towards e-governance by the rural citizen is measured.

Example 3
#When the Literacy on Health is 25 Awareness on ICT is 30.

Figure 9.5 describes the behavior pattern called "not using" among the rural citizens if there exists a low health literacy level "25" and weak literacy level on ICT "30". 33.5 fields has been found as the rate of utilization of e-governance services.

Example 4
#When the Literacy on Health is 25 and Awareness on ICT is 80.

Figure 9.6 describes that if the level of health literacy is low (25) and the awareness of ICT is very good (80) the rating is found to be 93.5 among the rural citizen for frequent practice of services. The value represents the pattern of "frequent" behaviour of the rural citizens for availing services under e-governance.

9.5 Discussion

The authors recognized the key factors that determine the use of e-Government services by rural residents. The researches have constructed the content analysis using fuzzy based system through the previous literature review. The authors have

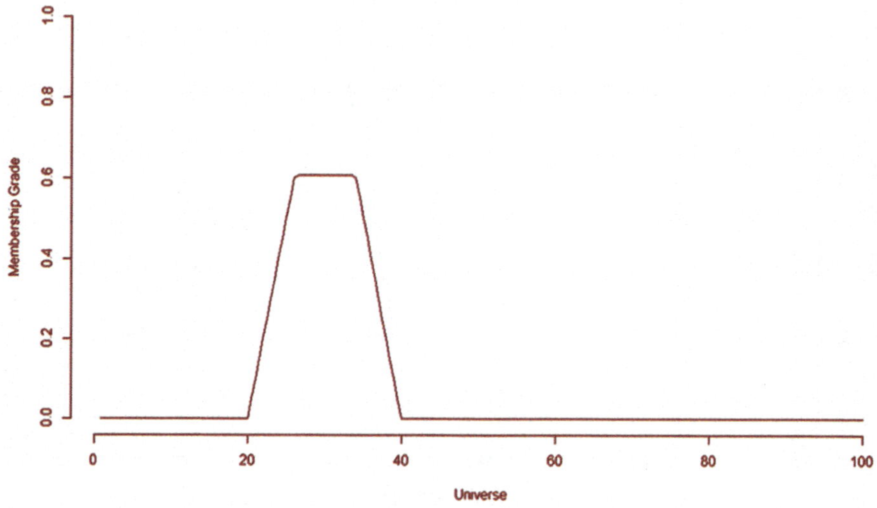

Fig. 9.5 Literacy on health is 25 awareness on ICT is 30

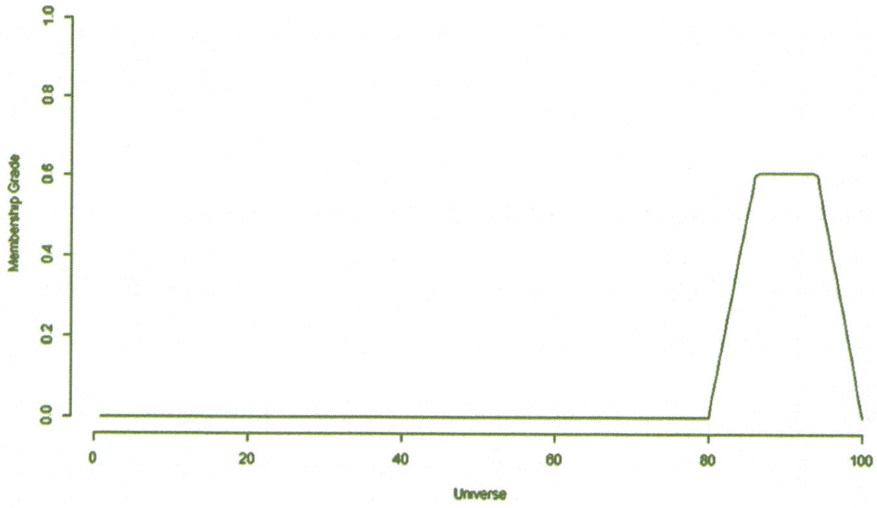

Fig. 9.6 Literacy on health is 25 and awareness on ICT is 80

constructed the 9 rules to evaluate the system of fuzzy logic that determines the 3 different sets of variables with its general properties. This study shows the overlap between the differences between the "medium" and "high" attributes of "health awareness". It is reasonable to assume that there is an overlap when health information falls in the 70–75 range, where users drop out and frequently use e-Government services. Researchers have learned that the use of ICT in education can help improve health knowledge in rural areas, thereby increasing their abilities and potential [48].

The present study have tested the fuzzy system with the 4 dissimilar samples through assigning the values to the different behaviour patterns of rural citizens. Information and communication technology (ICT) for rural development supports connecting rural people to access services and share information about work, agriculture and financial management. It also removes all control of intermediaries from schools and rural areas [49]. This shows that users of ICT can be better educated only if they have the lowermost literacy on health and ICT literacy as average [25].

The fuzzy systems were tested with little health literacy and feeble literacy on ICT as third example. Results obtained after the system of fuzzy logic are called unused tools of e-Government services in rural areas. This illustrates the need to work together to promote ICT literacy and health literacy. ICT information should be at the forefront because it provides more information, increases the level of knowledge, decreases the digital detachment and reinforces the country's financial level [50]. The system was tested with the last example by quoting little literacy on health and good knowledge on ICT. The present system produces the outcome based on the frequency of usage among the rural citizens on e-governance services. This proves that the literacy on health is found to be a development that gives a new knowledge and social patterns according to the variations in the time, place and society [51]. Hence the technology is said to be the basis for rural development and support the poor.

9.6 Conclusion and Future Work

A new initiative of Indian government to serve people and reinforce the association among the government and the people. The results of these projects provide step-by-step solutions to the problems rural people face through e-services. Literacy on ICT and the literacy on health are seen by governments and citizens as greater challenges in effectively managing e-Government services. This study predicts and analyses the real use of e-Government services by rural people by developing a fuzzy system. Fuzzy logic is tested with 6 different sets of rules to realize the system. The given variables as inputs are plotted to values and attributes that describe the act of the fuzzy system. The outcomes of the study supports the officials and policy makers determine access to government e-services for the rural population.

Current research is limited to two dimensions, ICT literacy and health literacy, to estimate the use of e-Government services by the rural population. By determining the dimensions of the fuzzy system and using various transformations with common properties. This system created by the authors could also examine the population of defendants by determining their income level, age, family members, and use of internet. This research presents an empirical study on the actual use of e-Government services. Future research will cover online marketing, online shopping, and digital marketing etc. can be done on other Internet services such as Therefore, this study clearly demonstrates the use of rural citizens' access models to e-Government services by creating a fuzzy system with fixed rules.

References

1. Abu-Shanab, E. (2014). Antecedents of trust in e-Government services: An empirical test in Jordan. *Transforming Goverment People Process Policy, 8*(4), 480–499.
2. Rossiter, D., & Bagdon, K. (1999). Embedding the acquisition of technological literacy: A case study. *Educational Technology & Society, 2*(4), 82–91.
3. Suri, P. K., Sushil. (2017). Introduction to E-governance. In *Strategic planning and implementation of e-governance.* Flexible Systems Management. Springer. https://doi.org/10.1007/978-981-10-2176-3_1
4. Bhatia, A., & Kiran, C. (2016). Rural development through E-governance initiatives in India. *IOSR Journal of Business and Management (IOSR-JBM), Special Issue—AETM'16*
5. Gupta, A., Gautam, S., & S. (2017). ICT for rural development: Opportunities and challenges. *International Journal of Information & Computation Technology, 7*(1), 13–23.
6. Tripathi, A. M., Singh, A. K., & Kumar, A. (2012). Information and communication technology for rural development. *International Journal on Computer Science and Engineering (IJCSE), 4*(5), 824–828.
7. Kak, S., & Gond, S. (2015). ICT for service delivery in Rural India –scope, challenges and present scenario. *OSR Journal of Computer Engineering (IOSR-JCE), 17*(6), 12–15.
8. Kamarudin, S., Omar, S.Z., Bolong, J., Osman, M.N., Mahamed, M. (2019). ICT development of community in rural areas. *International Journal of Academic research in Business & Social Sciences, 9*(9). E-ISSN: 2222–6990 © 2019 HRMARS.
9. Singh, U. (2019). e-governance implementation—A literature review analysis (March 6, 2019). *IJRAR—International Journal of Research and Analytical Reviews (IJRAR), 6*(1), 459–474. E-ISSN 2348–1269, P-ISSN 2349–5138. https://ssrn.com/abstract=3676985
10. Malodia, S., Dhir, A., Mishra, M., Bhatti, Z. A. (2021). Future of e-Government: An integrated conceptual framework. *Technological Forecasting and Social Change, 173*, 121102. https://doi.org/10.1016/j.techfore.2021.121102
11. Armstrong, L. J., Gandhi, N., & Lanjekar, K. (2012). Use of information and communication technology (ICT) tools by rural farmers in Ratnagiri district of Maharastra, India. In *2012 International Conference on Communication Systems and Network Technologies (CSNT)* (pp. 950–955). IEEE.
12. UN e-Government Survey. (2022). https://publicadministration.un.org/egovkb/en-us/Reports/UN-e-Government-Survey-2022
13. Arpaci, I., Yardimci Cetin, Y., & Turetken, O. (2015). Impact of perceived security on organizational adoption of smartphones. *Cyberpsychology Behavior and Social Networking, 18*(10), 602–608.
14. Turan, A. H. (2012). Internet shopping behavior of Turkish customers: Comparison of two competing models. *Journal of Theoretical and Applied Electronic Commerce Research, 7*(1), 77–93. ISSN 0718–1876.
15. Tan, K. S., Chong, S. C., Lin, B., & Eze, U. C. (2009). Internet-based ICT adoption: Evidence from Malaysian SMEs. *Indian Management Data Systems, 109*(2), 224–244.
16. Bhuvana, M., & Vasantha, S. (2017). A mediating effect of business correspondent model towards adopting mobile banking technology-a roadmap for achieving financial inclusion. *Journal of Advanced Research in Dynamical & Control Systems, 07*, 247–253.
17. Bhuvana, M., & Vasantha, S. (2019). Ascertaining the mediating effect of financial literacy for accessing mobile banking services to achieve financial inclusion. *International Journal of Recent Technology and Engineering (IJRTE), 7*(6S5).
18. Bhuvana, M., & Vasantha, S. (2020). Neural network machine learning techniques using R Studio for predicting the attitude of rural people for accessing E-governance services. *Journal of Advanced Research in Dynamical & Control Systems, 12*(2), 2552–2561.
19. *Predicting the Unpredictable with Emerging Technologies* (pp. 124–145).
20. Singh, M., & Sahu, G. P. (2018). Study of e-governance implementation: A literature review using classification approach. *International Journal of Electronic Governance, 10*(3), 237–260. https://doi.org/10.1504/IJEG.2018.095954

21. Zadeh, L. (1965). Fuzzy sets. *Information and Control, 8*, 338–353.
22. Carter, L., Yoon, V., Liu, D.: Analyzing e-Government design science artifacts: A systematic literature review. *International Journal of Information Management, 62*, 102430. https://doi.org/10.1016/j.ijinfomgt.2021.102430
23. Lewin, A. (2007). fuzzyFDR: Exact calculation of fuzzy decision rules for multiple testing. R package version 1.0. http://CRAN.R-project.org/package=fuzzyFDR
24. Schmidt, K. (2000). The problem with awareness. *Computer Supported Cooperative Work, 11*, 285–298.
25. Rahman, A. M., Mahfuz, M. U., Ahmed, K. M., & Rajatheva, R. M. A. P. (2005). ICT based sustainable rural business opportunities in developing countries: A wireless-networked RCP-RAP approach. *American Journal of Applied Science, 2*(8), 1256–1260.
26. Rozhkova, D., Rozhkova, N., & Blinova, U. (2021). Development of the e-Government in the context of the 2020 pandemics. In T. Antipova (Ed.), *Advances in digital science. ICADS 2021. Advances in Intelligent Systems and Computing* (vol. 1352). Springer. https://doi.org/10.1007/978-3-030-71782-7_41
27. Bhuvana, M., & Vasantha, S. (2019). An outlook of financial inclusion with mediating effect of direct benefit transfer in LPG subsidy towards actual usage of banking technology. *International Journal of Engineering and Advanced Technology (IJEAT), 8*(6S), 805–811.
28. Varun Kumar, M., & Venugopal, P. (2015). E-governance and rural development (A study specially focused on villages Of KatpadiTaluk, Vellore District of Tamil Nadu). *International Journal of Applied Engineering Research, 10*(92), 319–325.
29. Bhuvana, M., & Vasantha, S. (2020). Role of information and communication technology (ICT) for rural development through E-governance initiatives. *International Journal of Psychosocial Rehabilitation, 24*(8), 2705–2713.
30. Bhuvana, M., & Vasantha, S. (2020). Assessment of rural citizens satisfaction on the service quality of common service centers (CSCS) of e-governance. *Journal of Critical Reviews, 7*(5), 302–305.
31. Bhuvana, M., & Vasantha, S. (2020). Determinants of behavioral intention to access e-governance services by rural people with the mediating effect of information and communication (ICT) literacy. *Journal of Advanced Research in Dynamical & Control Systems, 12*(2), 176–187.
32. Bhuvana, M., and Vasantha, S. (2022). COVID-19 disaster healthcare management system in rural areas. Global Healthcare Disasters.
33. Bhuvana, M., Ramkumar, A., Neeraja, B. (2023). Machine learning technique for predicting the rural citizens' trust on using e-governance health care applications during COVID-19. In C. Gaie & M. Mehta (Eds.), *Recent advances in data and algorithms for e-Government. Artificial intelligence-enhanced software and systems engineering* (vol. 5). Springer. https://doi.org/10.1007/978-3-031-22408-9_6
34. Chiu, T. Y. (2015). *Micro-environment control by using fuzzy theory*. Master's Thesis, National Chung Hsing University, Taichung City, ROC (2015)
35. Davis, F. D. (1993). User acceptance of information technology: system characteristics, user perceptions and behavioral impacts. *International Journal Man-Machine Studies, 1993*(38), 475–487.
36. Dourish, P., & Bellotti, V. (1992). Awareness and coordination in shared workspaces. In *Proceedings of the 1992 ACM conference on Computer-supported cooperative work, CSCW '92* (pp. 107–114). ACM.
37. Gams, M., & Krivec, J. (2008). Demographic analysis of fertility using data mining tools. *Informatica, 32*, 147–156.
38. Government of India. (2015). India in Figures 2015. http://mospi.gov.in/
39. Gupta, S., & Xu, H. (2010). Examining the relative influence of risk and control on intention to adopt risky technologies. *Journal of Technology Management and Innovation, 5*(4), 22–37.
40. Gutwin, C., & Greenberg, S. (1999). The effects of workspace awareness support on the usability of real-time distributed groupware. *ACM Transactions on Computer-Human Interaction, 6*, 243–281.

41. Al-Kaseasbeh, H. M., Harada, Y., Naiemah, U., & binti Saraih. (2019). e-Government services assessment from the perspective of citizens interaction and satisfaction in Jordan: Pilot study. *International Journal of Research and Review, 6*(12), 50–60.
42. Jaeger, P. T. (2003). The endless wire: e-Government as global phenomenon. *Government Information Quarterly, 20*(4), 323–331.
43. Lindgren, I., Melin, U., & Sæbø, Ø. (2021). What is e-Government? In *Introducing a work system framework for understanding e-Government. communications of the association for information systems*, 48 pp. https://doi.org/10.17705/1CAIS.04842
44. Nyman-Metcalf, K. (2019). How to build e-governance in a digital society: the case of Estonia. *Revista Catalana de Dret Public 58*, 1–12. https://www.raco.cat/index.php/RCDP/article/download/357190/449147
45. Owo, N. J. (2010). Gender and development: Nigeria as a case study. *Journal of Sustainable Human Development Revision, 2*, 81–91.
46. Palvia, S. C. J., & Sarma, S. S. (2007). *e-Government and e-governance: definition/domain framework and status around the world*. Computer Society of India.
47. Philomine Roseline, T. N., Ganesan, C. T., & JM. (2015). A study of applications of fuzzy logic in various domains of agricultural sciences. *International Journal of Computer Applications, 0975–8887*, 15–18.
48. Gupta, P. (2017). Application of fuzzy logic in daily life. *International Journal of Advanced Research in Computer Science., 8*(5), 1795–1800.
49. Petare, P.A.: Information and communication technology for rural development. *ASM's International E-Journal of Ongoing Research in Management and IT*, INCON13-IT-048, 1–8.
50. Rao, S. (2004). Role of ICTs in India's rural community information systems. *Info, 6*(4), 261–269.
51. Makkar, R. (2018). Application of fuzzy logic: A literature review. *International Journal of Statistics and Applied Mathematics, 3*(1), 357–359.

Chapter 10
Enhancing the Efficiency and the Security of e-Government: The French Case Study of Human Resources Applications

Christophe Gaie

Abstract This chapter explores advancements in e-Government by introducing innovative architectures: a data-based model and a procedure-based model. The data-based approach emphasizes secure data exchange through APIs, addressing the balance between data sharing and privacy protection. It leverages an API catalog, enrollment module, and manager to facilitate efficient inter-ministerial communication. The literature review underlines the importance of IT security, administrative efficiency, and user satisfaction in e-Government, providing insights into existing models. The procedure-based architecture focuses on pooling services between IT departments to accelerate e-Government adoption. It allows multiple departments to use a single service, enhancing overall efficiency. The proposed frameworks offer practical solutions to challenges in data exchange, security, and administrative efficiency. The paper details the operational steps in both architectures, emphasizing security measures in the data-based model and efficiency gains through shared services in the procedure-based model.

Keywords e-Government · Inter-ministerial architectures · Data-based model · Procedure-based model · API exchange · IT security · Administrative efficiency · User Satisfaction

C. Gaie (✉)
Chief of Engineering and Innovation Department, French Prime Minister Services, 75007 Paris, France
e-mail: christophe.gaie@gmail.com

C. Gaie and M. Mehta (eds.), *Transforming Public Services—Combining Data and Algorithms to Fulfil Citizen's Expectations*, Intelligent Systems Reference Library 252, https://doi.org/10.1007/978-3-031-55575-6_10

10.1 Introduction

During the last decades, public services have flourished and modernized in multiple domains. Citizens are able to declare their personal income online, to obtain a driving licence, to require from a subsidy online, etc. As this modernization is useful, digitalizing public services is only the first step toward a deeper change. Indeed, the adoption of Information and Communications Technology for public services offers a remarkable opportunity to reinvent government and define e-Government [1].

It is now possible to combine data to provide new services that may not require any intervention of citizens: withholding tax on personal income, automatic subsidy payment, parking infraction detection, etc. Thus, there is a huge interest in developing data exchanges between government administrations [2]. This raises the interest to refine the well known « four phases of e-Government model » of Baum and Di Maio [3] by introducing a phase of automatization and forecasting that relies on microservices and API.

As a matter of fact, the evolution of technology enables to take better advantage of data, to combine them and offer a better experience to citizens. Artificial intelligence can forecast citizen needs and anticipate their behaviours. For instance, when a hospital records a birth, the parents should not have to move to the town hall, re-validate the administrative situation, send the information to the social insurance agency, inform the tax administration, send information to the family allowance agency, etc. A key objective of eGovernment is to remove this burden and automatize these procedures, as illustrated in Figure 10.1.

While this new trend offers many opportunities, it also requires complying to the regulations defined by the authorities. To be more specific, a legal framework was defined to ensure data protection and protect people's right to privacy. Indeed, the European Commission established the General Data Protection Regulation (GDPR)

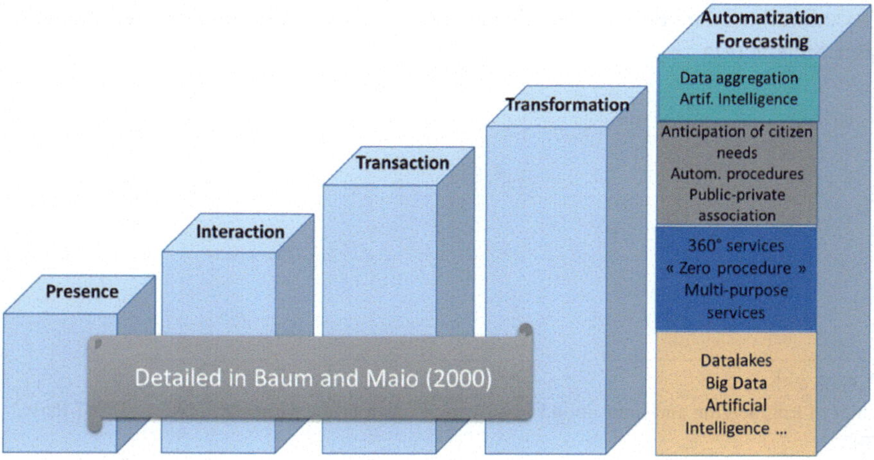

Fig. 10.1 Extension of the Gartner's Four Phases of e-Government Model from Baum and Maio

2016/679 as the laws that any European Information Technology (IT) application has to fulfill.

As a consequence, each government administration has to implement security measures to protect the private data of citizens. This protection tends to restrict data exchange to ensure a better protection. How to define a secured architecture that offers a tradeoff between the aspirations of developing data exchange to ensure e-Government and the desire to protect personal data? Which security modules offer the relevant protection while enabling secured communication between public services?

The communication between two endpoints is usually secured through a secured channel and a dedicated exchange system. This aims to benefit from the widespread use and maturity of Application Programming Interfaces (API) whose description was notably provided by Gottschalk et al. [4]. Implementing this channel requires ensuring the discoverability of API, the traceability of exchanges, the protection of personal data, etc. In the context of Human Resource Information System (HRIS), there is a need to collect data from an Enterprise Resource Planning (ERP) whose data storage is specific, and transform it so that it facilitates reutilization and fulfils RESTful implementations as illustrated in Fig. 10.2.

The data-based architecture described above offers a secured architecture with an exclusive data management application (i.e. the HRIS). The classic communication that relies on exchange of data through .csv files is replaced by an automatized API architecture, but it is still required to develop procedures in each government administration.

When a common procedure exists (for instance: professional assessment, skill management, training and career development, etc.), the same procedure may be implemented in thousands of administrative entities (each ministry, agency, university, hospital, etc.). Thus, a new and innovative approach consists of implementing

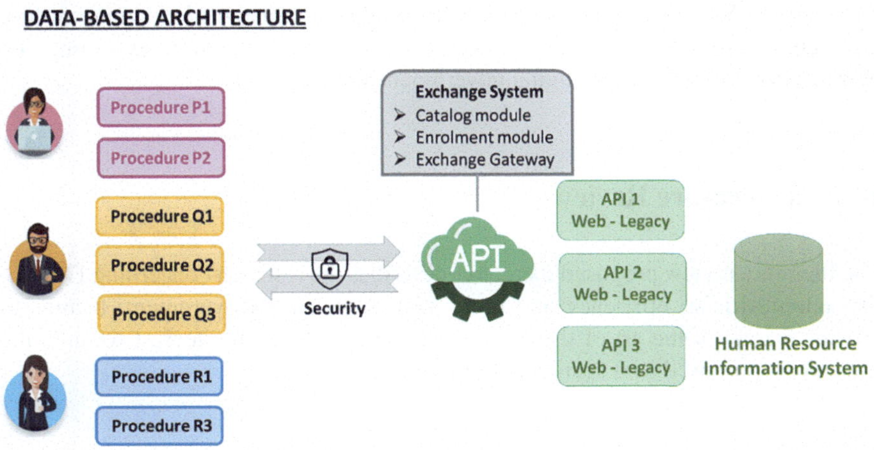

Fig. 10.2 Illustration of the architecture that optimizes the reuse of data

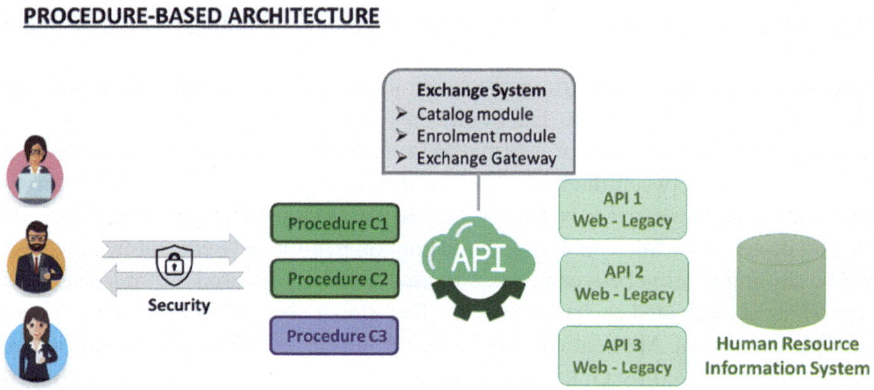

Fig. 10.3 Illustration of the architecture that optimize the reuse of procedures

common procedures and sharing them directly with government administrations, as described in Fig. 10.3.

This innovative approach will be described in the current chapter with an objective to highlight its benefits in terms of IT security, administrative efficiency and user satisfaction.

The chapter is organized as follows: in Sect. 10.2, the literature Review, delves into existing research on IT security, administrative efficiency, and user satisfaction within the e-Government landscape. Then, Sect. 10.3, details the Data-Based Inter-ministerial Architecture and the proposed data-centric model, covering secure data exchange, API discovery, and operational steps. Thereafter, Sect. 10.4, describes the Procedure-Based Inter-ministerial Architecture. This architecture outlines the mutualization of services among IT departments to expedite e-Government adoption. Finally, Sect. 10.5 provides a conclusion that summarizes key contributions, highlighting the benefits of both architectures, and offering perspectives for ongoing optimization in the realm of digital government services.

10.2 Literature Review

The literature review proposed thereafter describes the research concerning IT security, administrative efficiency and user satisfaction for e-Government. It aims to provide a state-of-the art of these domains and to outline the interest towards the framework proposed next sections of this chapter.

10.2.1 Ensuring IT Security of e-Government Applications

First of all, the literature describes the importance of ensuring security in e-Government procedures reveals fundamental. As a matter of fact, the lack of security reduces the confidence in public services and the adoption of electronic processes [5]. The survey findings identified multiple limitations to security in public IT systems, such as deficiencies concerning user privacy, trust in the technology, guarantee of confidentiality, data protection, etc. To improve the situation, the authors propose to implement an e-Government adoption model that is unfortunately not detailed. On the contrary, the current chapter proposes a concrete and detailed framework.

A useful framework was proposed to manage security in IT project by Sangi et al. [6]. The method combines System Development Life Cycle Security and IT Project Management Models to integrate security at each level of IT project management. The authors indicate that Confidentiality, Integrity and Availability can be managed by five processes: Initiation, Planning, Executing, Monitoring and Controlling, Closing. Whereas this approach is promising, it should be further developed to offer a practical implementation. Unlike numerous theoretical approaches, an analysis of government agencies' website brings a lot of lessons. For instance, Elisa proposed a detailed diagnostic of institution websites in Tanzania [7]. This verification underlined that more than a half of websites presented a high-severity vulnerability (SQL injection or cross site scripting-XSS) while two-thirds presented medium-severity weaknesses (Cross site request forgery or Denial of Service). Whereas the research is instructive, recommendations to improve security reveal insufficient, as the author only proposes to cypher passwords and validate inputs before further computations.

A novel approach to combine security of e-Government services and ease of development consists in implementing a Cloudlet based security trust model [8]. A Cloudlet is an architectural component that combines multiple trusted data centers connected to the Interne; this component may be used by any nearby mobile devices. The model proposed leans on a governmental Cloudlet that acts as a trusted third party between web services. The proposed model provides interesting performance in the context of SOAP messages cyphered with triple DES and RSA algorithms.

Enforcing security of procedures and protecting the privacy of personal data are key factors in the emergence of e-Government [9]. The researchers propose a conceptual model based on latent variables: perceived trust, privacy, security, risk, etc. Their analysis based on survey lead in Zimbabwe and Zambia proved that confidence is mainly subjective. Thus, the authors recommend that the government should work to reduce the uncertainty towards the usage of dematerialized public services. Whereas these conclusions are useful as they insist on the final user's perception, the authors do not provide concrete proposals to manage security and improve e-Government adoption. Another model was proposed to examine the role of the perceived risk and security to determine the level of trust in e-Government [10]. The results obtained through a survey performed in Poland confirm that using a security variable encourages citizens to adopt the web solutions offered by administrations. Both of these research papers underline the importance of data securization to foster the adoption of e-Government. This teaching is clearly taken into consideration in the framework proposed in the current chapter.

Establishing a high level of confidence in the usage of public services requires ensuring data security and privacy for every e-Government application [11]. Saudi Arabia presents a certain number of security weaknesses, such as the lack of monitoring and accountability of governmental or the insufficient skills of IT manpower. The interviews performed in the survey indicate that Saudi Arabia made efforts to secure data and provide an adequate fiber network to facilitate the adoption of public services. These actions should be supplemented by audits to certify the current level of security and implement the proposed measures to boost the confidence of citizens. This chapter considers the securization of public services as a core objective of the framework. This will contribute to increase the confidence in e-Government applications.

The use of Artificial Intelligence enhances cybersecurity of government administrations [12]. As a matter of fact, the understanding of normal behaviors of public services enables to predict whether an attack is about to occur and to establish immediate protection measures. The research pursues three objectives: 1—determine the type and characteristics of cyber-attacks; 2—establish vulnerabilities to predict potential attacks on the system; and 3—design an artificial intelligent to identify attacks and prevent future damage. The paper also details the proposed roadmap to deploy the mechanism in the Philippine Academe. These results will offer a deeper understanding of the framework effectiveness. The use of AI offers future perspectives to optimize the proposed framework. As a matter of fact, AI can play a crucial role in the valorization of data held by Human Resource services.

10.2.2 Improving Administrative Efficiency and Performance Through the Adoption e-Government

Improving the efficiency and performance of public services is a key concern for both civil servants and researchers. Understandably, the use of technology is assumed to improve the functioning of government services, but this assumption needs to be verified. The research of Chen and Aklikokou provides valuable information on the benefits of technology and e-Government to improve government effectiveness [13]. The authors compare the development of e-Government in 191 countries using the Public Service Bargains (PSB) and compare them notably according to the level of professionalism and autonomy of civil servants. The results indicate that the lower the bureaucratic organization is, the higher the development of e-Government as well as public services effectiveness. The current chapter is inspired by the functioning of the French Administration,[1] which is identified by the authors as a leading country for e-Government. Thus, the proposed framework may be easily applied in similar countries.

[1] The author is a French civil servant since 2009.

The performance of public services may be measured through multiple indicators. For instance, Ma and Zheng evaluate the performance with eight parameters (Overall e-govt satisfaction, E-information, E-service satisfaction, etc.) and compute their correlation by considering a satisfaction survey of over 28,000 citizens across 32 European countries [14]. The research points out that e-Government performance indicators are positively coupled with citizen satisfaction and perceived benefits. While this conclusion is valuable, it does not indicate how to improve the performance of e-Government. This dimension will be tackled in the current chapter.

It was also proved that the adoption of e-Government largely contributes to improve the efficiency of public services. For example, Jauhari et al. proposed a model based on the structural equation modelling (SEM) to analyze the contribution of multiple variable illustrating e-Government and bureaucratic reform to enhance organization performance [15]. Through the use case of public services in Indonesia, the authors outline that implementing good governance based on bureaucratic reform and e-Government increases the performance of the public organizations in Indonesia. They also provide some interesting recommendations to increase the organization's performance. While the results are interesting, they do not provide a method to optimize the efficiency of public services and their recommendations are specific.

The research towards e-Government has flourished in the last decades and initiatives tend to lack of interoperability. As a matter of fact, each government administration tries to optimize its own services without natively considering other government services. The solutions proposed also often lack of a clear organization in terms of data acquisition, network functioning and application development. Thus, Qi et al. proposed a very innovative approach that structures e-Government using a three-layer framework that combines data, networks and applications [16]. While this approach is valuable and necessary, it does not illustrate concrete use cases.

The comparison and emulation between comparable countries is also a catalyst to improve the impact of e-Government adoption on the effectiveness of public services. Doran et al. describe the different factors that contribute to government efficiency and propose a framework for comparing the maturity of European countries [17]. To this end, the authors define three indicators: the online service index (OSI), the telecommunication infrastructure index (TII), and the human capital index (HCI). Then, they combine them into the e-Government development index (EGDI) that measures government effectiveness. This simple approach is very useful to outline the dimension to improve in a given country. However, it does not explain how to do it and requires exploration.

10.2.3 Enhancing User Satisfaction by Developing e-Government

Another key objective of this state-of-the art is to highlight the mutual effects between the increase of user satisfaction and the adoption of e-Government. As a matter of fact, each public service user is more likely to interact using a digital mean with the Administration when the service offers an added value compared to the classic face-to-face procedure. This explains why proposing modern and digitalized government services contributes to the satisfaction of public services. In the meantime, the simplicity, swiftness, and creation of user value offered by digitalized procedures contributes to the adoption of dematerialized procedures and the success of e-Government.

The concept of "Public Value" was defined a long time ago by Moore in 1995 [18] and refers to the benefits of public activities, with many similarities to the value created in the private sector. While this concept is rather old, its concrete implementation in public processes still has to progress. A key factor of improvement consists in shifting the focus from the government perspective to the user consideration. In [19], MacLean and Titah propose a valuable categorization of impacts (capability, interactions, orientation, distribution) according to the role held by a public service user (citizen, client, obligatee, taxpayer). This approach should be taken into consideration for each new service proposed by government organizations.

Valle-Cruz et al. proved the benefits of providing public value to accelerate the adoption of e-Government in [20]. Their research leans on a linear regression model for the metropolitan area of the Toluca Valley (Mexico). It identifies various benefits that plays a positive effect towards the adoption of digital public services: anti-corruption strategies, open data, access to information and data privacy. A limitation of the proposed approach consists of relying on a pre-identification of public value effects (hypotheses) before their potential validation by the model (verification).

A comprehensive review of the public value yielded by e-Government was proposed by Twizeyimana and Andersson in [21]. This research highlights six dimensions of public values: 1—public services; 2—administrative efficiency; 3—Open Government (OG) capabilities; 4—ethical behaviour and professionalism; 5—trust and confidence in government; and 6—social value and well-being. It also proposes a new framework based on three augmented dimensions: Improved Administrative Efficiency, Improved Public Services, and Improved Social Values. While the research is valuable, it does not explain how to implement it concretely.

Interestingly, Maragno et al. proposed a conceptual framework to enhance public values enhancement thanks to e-Government projects [22] that take advantage of the augmented dimensions proposed by Twizeyimana and Andersson. Above these expected benefits, the authors also describe the perspectives depending on the scope of impact and the conditions in which the project is led. This research offers an interesting state-of-the-art concerning public value creation for e-Government. However, the model is not concretely detailed to compare different implementations or identify possible improvements.

A recent research highlights the dimensions that contribute to the adoption of eGovernment [23]. Mensah and Zeng propose a concrete the Public Value of the e-Government Adoption Model (PVEGAM) that combines multiple dimensions and identifies their correlation with the adoption. Using a sample of 517 data generated from Chinese citizens, the results indicate that efficiency and quality of information are the factors of highest importance. Again, this information is very useful but does not indicate how to implement it concretely.

10.3 Data-Based Inter-Ministerial Architecture

The data-based architecture of Human Resource Information System relies on the modules described in Fig. 10.4 and detailed.

- **A secured channel**: this aims to ensure cyphered exchanges between organizations. In France, this channel is dedicated to government administrations and is separated from the common internet network [24, 25]. The inter-ministerial network of the State (RIE[2]) relies on a single operator and an inter-ministerial collaboration. It offers a mutualized infrastructure that secures efficiently the IT infrastructure of the French State and streamlines data exchanges for public services.
- **An API Catalog**: this module is dedicated to the discoverability of public services (see [26], Chap. 3). As a matter of fact, increasing the visibility of available data and services is necessary to offer improved public services. This provides value to the preliminary stage of modernizing Legacy systems through API [27, 28]. In France, the API Catalog for public services is https://api.gouv.fr/.
- **An API Enrolment Module**: the enrolment, also known as registration, was not originally implemented in REST architectures [29]. However, the requirement to protect sensitive information concerning organizations or people leads to a restriction of the exchange of information to authorized entities. For example, this aims to ensure that only some specific public services may obtain fiscal information concerning a person and may utilize them for a dedicated purpose. In France, the API Enrolment Module for public services is https://datapass.api.gouv.fr/.
- **An API Manager**: this module aims to manage the overall API lifecycle within organizations. Thus, it enables to plan, design, implement, test, deploy, operate, and version of APIs in a simple, secure, and efficient manner to optimize the usage of data. This platform grants the ability to publish APIs and enable their consumption by authorized systems. Among the available open-source and proprietary solutions detailed in [30], the French Ministry of Finances deployed an Axway gateway entitled https://piste.gouv.fr/. This platform reveals to be robust and offers

[2] RIE: Réseau Interministériel de l'Etat.

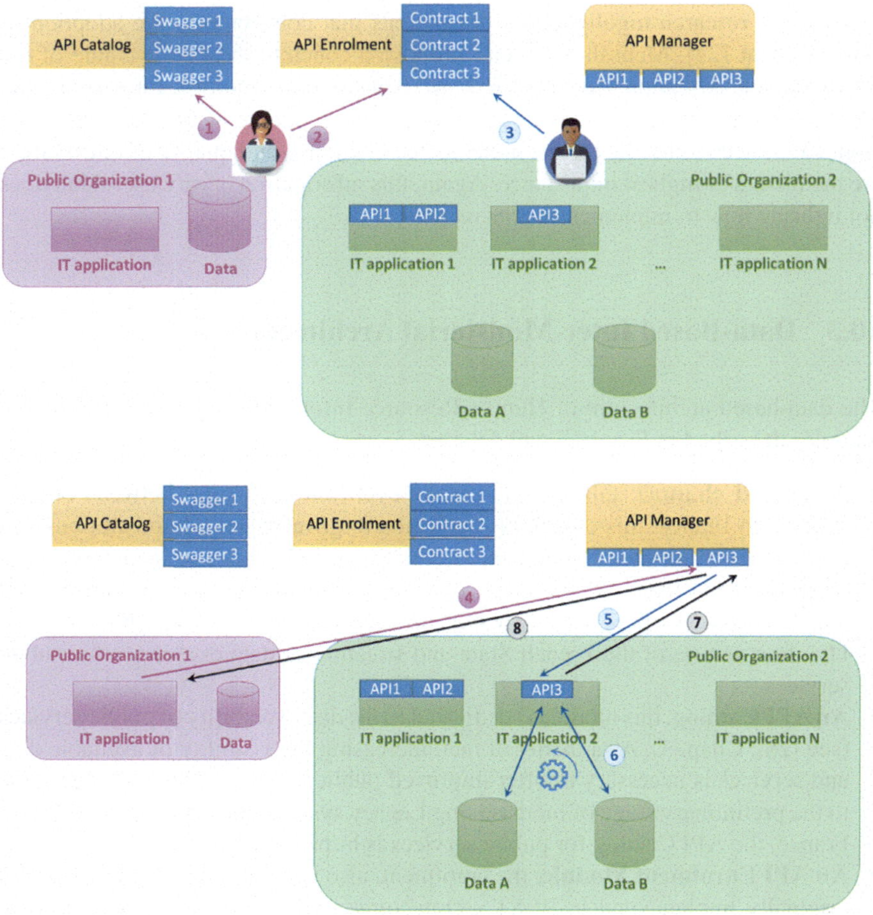

Fig. 10.4 Functioning of the data-based inter-ministerial architecture

an interesting mutualization for the multiple organizations that interacts with the Ministry.

As the architecture was detailed and explained, the operations below describe how an organization 1 can obtain information from an organization 2.

- **Step 1**: the public organization 1 wants to construct a new service that requires information from the public organization 2. It doesn't know exactly which information to use and consults the available API as well as their description (Swagger, i.e. OpenAPI Specification) published in the common API Catalog. After an expertise, the organization 1 decides to request access to data exposed through API 3.
- **Step 2**: the organization 1 fulfills in the Enrolment module the form that may offer access to API 3. It describes its use case, the legal basis that justifies the usage

of this data and indicates the different persons involved in the electronic procedure. After potential interactions between the organizations through the module messaging service, the agreement is granted and stored in case of a later audit by the national data protection authorities (in France, this is the Commission nationale de l'informatique et des libertés)

- **Step 3**: the organization 2 examines the access request and decides whether an authorization has to be granted to organization 1. The two organizations may exchange information through the secured message service offered by the API Enrolment module. Then a decision is made concerning the access to API 3. Providing that the answer is positive, a private key of usage is delivered to organization 2 using a secured method
- **Step 4**: the organization 1 transmits an HTTPS request to API 3 exposed on the API Gateway, using the private key that it possesses. The API Gateway verifies the authorizations granted to organization 1. If the verification is positive, then the Gateway transmits the request parameters to the application in order to compute and obtain data that corresponds to the request.
- **Steps 5 and 6**: the API 3 receives the request from the API Gateway and applies the process described in the API documentation. This may require gathering data from multiple sources and perform calculation. For example, the requiring organization will probably find it useful to obtain simultaneously data concerning skills as well as data trainings. This enables to perform only 1 request when information is linked.
- **Steps 7 and 8**: the results obtained by API 3 are transmitted to the API Gateway that sends the final response to organization 1.

This approach was implemented in France to ensure the connection between multiple government organizations [2]. An additional layer of security can also be implemented as proposed in [31]. While the approach relies on a secured and modular functioning, it only shares raw data and requires building IT applications for each government organization. The concrete need for government services is not to obtain data and re-recreate an existing procedure such as professional assessment, skill management or teaching management. The requirement is to use an existing procedure with some minimal configurations.

Thus, the following section proposes a framework that detail how to implement (by a unique inter-ministerial structure) a procedure-based solution that accelerates the adoption of IT services by multiple IT departments.

10.4 Procedure-Based Inter-Ministerial Architecture

The procedure-based architecture of Human Resource Information System aims to mutualize services between different IT departments.

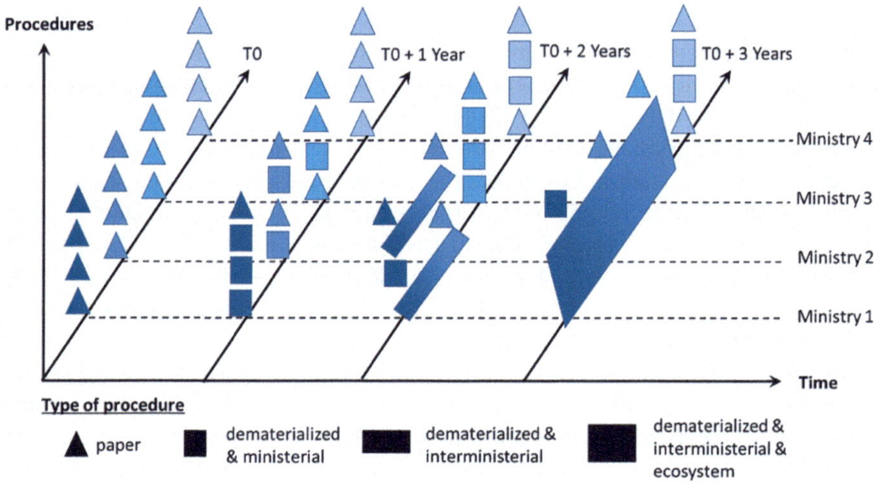

Fig. 10.5 Efficiency gain obtained by a procedure-based architecture

As indicated in Fig. 10.5, it accelerates the adoption of e-Government services and increases the global efficiency since 1 service may be used by many IT departments. This offers the opportunity to create better software with mutualized human resources. Indeed, when a unique service proposes a digital procedure, a part of the efficiency gain may be used to offer a better service. For example, the French professional evaluation of civil servants is used by more than fifty government services.

To be more specific, the French IT Human Resource Service (entitled "CISIRH" in French) offers multiple services to government services from various ministries (rural affairs, ecology, health and social affairs, education, etc.). Concretely, the system is connected to a core Human Resource Information System (HRIS) entitled RenoiRH[3] that enables to manage worker time, attendance, training, payroll, etc. This system relies on a large and comprehensive private software that is adjusted to the French.

Administration rules. Today, the French HRIS manages 400 000 civil servants for approximately twenty business processes. While it covers a very large scope of HR procedures, new services are emerging to cover specific requirements, such as careers, evaluation, skills, mobilities, etc. Figure 10.6 illustrates this functioning.

The functioning is different from the previous approach and offers the following advantages:

- **Mutualization**: there are multiple services that can be mutualized. For example, the evaluation of civil servants follows a unique process in each Ministry. Thus, the French administration offers a unique web service entitled ESTEVE[4] that enables

[3] RenoiRH is inspired from Auguste Renoir who was a famous french painter.

[4] Etsève is inspired from Maurice Estève who was a famous french painter.

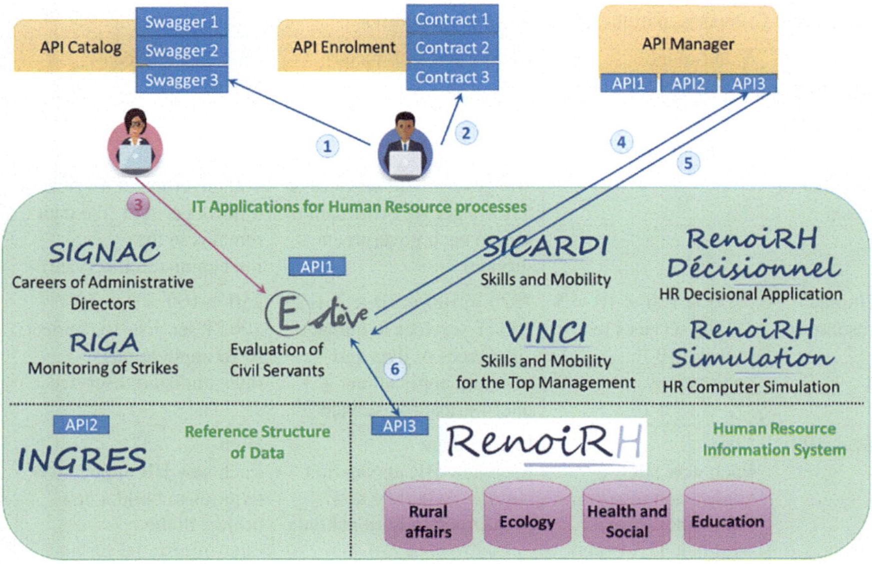

Fig. 10.6 Functioning of the procedure-based architecture

to evaluate the performance of civil servants every year. As this service is provided by the CISIRH, it replaces former evaluation systems within Ministries and provides a better service as there are more resources to maintain it in operational condition.

- **Information Reuse**: the proximity of multiple HR solutions offers the opportunity to exchange information efficiently. As there are multiple IT software handled by the same government organization, there is a higher opportunity to exchange data between applications. For instance, it is possible for ESTEVE to obtain information concerning the personal situation of civil servants through RenoiRH by using API3. Similarly, it is possible to obtain information concerning the label of any grade or skill by sharing information with INGRES[5] that is the application in charge of reference data.
- **Security**: as the data remains within a single organization, it is easier to protect the system from external attacks. Each request (step 3) from a partner is transmitted through the Inter-ministerial Network of the State [24]. This network ensures that the protection of the communication through secure internet browsing services [32]. The authentication relies on a federated identity process called AgentConnect that allows a user to log in with the identity held in its own organization.

A comparison of the efficiency of the different architecture highlights the benefits of mutualization. To this end, let's suppose that 100 government services want to connect to 10 sources of data or 10 webservices (Table 10.1).

[5] Ingres is inspired from Jean-Auguste-Dominique Ingres who was a famous french painter.

Table 10.1 Comparison of the efficiency of architectures

	Stand-alone architecture	Data-based architecture	Procedure-based architecture
Sources of data	100 × 10 (each service is internal)	10	10
Number of flows	0	100 government services × 5 flows per API (the data is sent to each government service)	100 government services × 4 flows per API (the data remains in the inter-ministerial service)
Human resources	5000 = 100 × 10 × 5 100 IT services to manage 10 applications	3020 = 100 × 10 × 3 + 20 100 IT services to manage 10 sources of data and maintain applications + 1 inter-ministerial service	250 = 100 × 2 + 50 100 IT services to connect to 10 applications + 1 inter-ministerial service
Scalability	Low Each new HR application requires to build a new project with new IT workers	Intermediate Each new HR application requires to build a new project with additional data	High Each new HR application requires to build a new project in the inter-ministerial service

Furthermore, we can compare the classic and modern architecture according to different potential cyberattacks (Table 10.2).

A key approach to secure the system consists in strengthening both the inter-ministerial platform and the network to connect it to external government services. To this end, the secured channel should combine multiple technologies such as IPsec

Table 10.2 Comparison of the architecture in terms of security

	Data-based architecture	Procedure-based architecture
Unauthorized access	An attacker can obtain data by using an authentication in both IT systems	An attacker can obtain data only by using an authentication of the IT HR service but it can obtain rights to multiple IT software
Hijacking the API manager	There are numerous partners of the API Manager to supervise	The IT HR service is the only organization to exchange information with the API Manager
Data theft	An attacker can obtain data by targeting the databases of both IT systems	An attacker can obtain data only by targeting the databases of the IT HR service
Distributed denial of service	Attackers may attack each IT application separately. It is more difficult to protect, but an attack is more restricted	Attackers may attack the whole system's IT application separately. It is easier to protect but an attack has a greater impact

(Internet Protocol Security) and TLS (Transport Layer Security). It is also possible to introduce multi-layer security with ML-IPsec [33] or MTLS [34].

10.5 Conclusion

In this chapter, the author presented novel inter-ministerial architectures aimed at bolstering the efficiency and security of e-Government services. The foundational elements of these architectures are expounded through a theoretical lens, delineating the distinctive attributes of both the data-based and procedure-based models. The data-centric model increases the secure exchange of information via APIs, while preserving an equilibrium between data accessibility and privacy safeguards. This is facilitated through the deployment of an API catalog, enrollment module, and API manager.

Afterwards, an improved approach was proposed and relies on Procedure-Based Inter-ministerial Architecture. It provides the strategic mutualization of services between IT departments to expedite the adoption of e-Government. This approach centralizes services, enhancing overall efficiency and resource utilization.

The proposed frameworks go beyond theoretical constructs, providing implementable models for optimizing digital government services. The outlined steps offer a pragmatic roadmap for policymakers, IT professionals, and researchers to enhance service delivery in the dynamic landscape of public administration.

Further research will tackle the implementation of these frameworks in other professional contexts, such as the procedures concerning identity verification, granting social aids, etc. The work presented in this chapter can be extended in terms of cybersecurity as well as specific configurations aiming to fulfill the different expectations of organizations.

Disclaimer This disclaimer informs readers that the views, thoughts, and opinions expressed in the text belong solely to the author, and not necessarily to the author's employer, organization, committee or other group or individual.

References

1. Silcock, R. (2001). What is e-Government. *Parliamentary Affairs, 54*(1), 88–101. https://doi.org/10.1093/pa/54.1.88.
2. Gaie, C. (2022). An API-intermediation system to facilitate data circulation for public services: The French case study. *International Journal of Computational Systems Engineering (IJCSYSE), 6*(4), 2021. https://doi.org/10.1504/IJCSYSE.2021.120292.
3. Baum, C., & Di Maio, A. (2000). Gartner's four phases of e-Government model. *Gartner Group, 12*, 207–214.
4. Gottschalk, K., Graham, S., Kreger, H., & Snell, J. (2002). Introduction to Web services architecture. *IBM Systems Journal, 41*(2), 170–177. https://doi.org/10.1147/sj.412.0170.

5. Alharbi, N., Papadaki, M., & Dowland, P. (2014).Security Factors Influencing End Users? Adoption of e-Government. *Journal of Internet Technology and Secured Transactions (JITST) 3*(4). https://infonomics-society.org/wp-content/uploads/jitst/published-papers/volume-3-2014/Security-Factors-Influencing-End-Users-Adoption-of-e-Government.pdf.
6. Sangi, S., Ilkan, M., & Tokgöz, H. (2017). Incorporating information security into IT project management (A Proposed Framework)." (2017). *International Journal of Computing, Communication and Instrumentation Engineering 4*(1). https://doi.org/10.15242/IJCCIE.ER0317112.
7. Nnko, E., & Noe. (2017). Usability, accessibility and web security assessment of e-Government websites in Tanzania. *International Journal of Computer Applications., 164*, 42–48. https://doi.org/10.5120/ijca2017913632.
8. Al-Shargabi, B. A. S. S. A. M., AlJawarneh, S. H. A. D. I., Hayajneh, S. M. (2020) A cloudlet based security and trust model for e-Government web services. *J Theor Appl Inf Technol 98*, 27–37. https://www.jatit.org/volumes/Vol98No1/3Vol98No1.pdf.
9. Munyoka, W., & Maharaj, M. (2019). Privacy, security, trust, risk and optimism bias in e-Government use: The case of two Southern African Development Community countries. *SA Journal of Information Management, 21*(1), 9. https://doi.org/10.4102/sajim.v21i1.983.
10. Ejdys, J., Ginevicius, R., Rozsa, Z., & Janoskova, K. (2019). The role of perceived risk and security level in building trust in e-Government solutions. *E + M Ekonomie A Management, 22*(3), 220–235. https://doi.org/10.15240/tul/001/2019-3-014.
11. Alharbi, A. S., Halikias, G., Rajarajan, M., et al. (2021). A review of effectiveness of Saudi e-Government data security management. *International Journal of information Tecnology, 13*, 573–579. https://doi.org/10.1007/s41870-021-00611-3.
12. Aquino, M. F. M., Noroña, M. I. (2021). Enhancing cyber security in the Philippine academe: A risk-based it project assessment approach. In *Proceedings of the international conference on industrial engineering and operations management* (pp. 5166–5179). http://www.ieomsociety.org/singapore2021/papers/878.pdf.
13. Chen, L., & Aklikokou, A. K. (2019). Relating e-Government development to government effectiveness and control of corruption: A cluster analysis. *Journal of Chinese Governance*. https://doi.org/10.1080/23812346.2019.1698693.
14. Ma, L., & Zheng, Y. (2017). National e-Government performance and citizen satisfaction: A multilevel analysis across European countries. *International Review of Administrative Sciences, 85*(3), 506–526. https://doi.org/10.1177/0020852317703691.
15. Jauhari, A., Abd, M. M., Basri, H., & Djalil, M. (2020). Are e-Government and bureaucratic reform promoting good governance towards a better performance of public organization? *Calitatea, 21*(175), 25–30.
16. Qi, M., Wang, J., & Lv, Z. (2021). Using the internet of things e-Government platform to optimize the administrative management mode. *Wirel Commun Mob Comput 2021*. https://doi.org/10.1155/2021/2224957.
17. Doran, N. M., Puiu, S., Bădîrcea, R. M., Pirtea, M. G., Doran, M. D., Ciobanu, G., & Mihit, L. D. (2023). e-Government development—a key factor in government administration effectiveness in the European Union. *Electronics, 12*(3), 641. https://doi.org/10.3390/electronics12030641
18. Moore, M. H. (1995). *Creating public value: Strategic management in government*. Harvard University Press.
19. MacLean, D., & Titah, R. (2022). A systematic literature review of empirical research on the impacts of e-Government: a public value perspective. *Public Admin Rev, 82*, 23–38. https://doi.org/10.1111/puar.13413
20. Valle-Cruz, D. (2019). Public value of e-Government services through emerging technologies. *International Journal of Public Sector Management, 32*(5), 530–545. https://doi.org/10.1108/IJPSM-03-2018-0072
21. Twizeyimana, J. D., & Andersson, A. (2019). The public value of e-Government—a literature review. *Government Information Quarterly, 36*, 167–178. https://doi.org/10.1016/j.giq.2019.01.001

22. Maragno, G., Gastaldi, L., & Corso, M. (2021). e-Government for Public Values creation: a systematic literature review. In *DG.O2021: The 22nd Annual International Conference on Digital Government Research (DG.O'21)* (pp. 386–397). Association for Computing Machinery. https://doi.org/10.1145/3463677.3463692.
23. Mensah, I. K., Zeng, G., & Mwakapesa, D. S. (2022). Understanding the drivers of the public value of e-Government: Validation of a public value e-Government adoption model. *Frontiers in Psychology, 13*(13), 962615. https://doi.org/10.3389/fpsyg.2022.962615.
24. Chausson, C. France mutualises its network infrastructure to secure its IT Published on: 20/07/2015. https://joinup.ec.europa.eu/collection/egovernment/document/france-mutualises-its-network-infrastructure-secure-its-it.
25. Lambert, G., Fontaine, B., Monneret, M., & Madani, R. M. (2019). How to build an innovative C2 system supporting individual-centric emergency needs? In *Proceedings of the 16th ISCRAM Conference—València, Spain.* http://idl.iscram.org/files/guillaumelambert/2019/1907_GuillaumeLambert_etal2019.pdf.
26. Subramanian, H., & Raj, P. (2019). *Hands-on RESTful web API design patterns and best practices: design, develop, and deploy highly adaptable, scalable, and secure RESTful web APIs.* https://www.overdrive.com/search?q=8D54DEC5-7EA6-4B11-97EB-78FF5EDF3A3C.
27. Flaishans, J., Galvin, M., Ignatius, A., Kuan, C., Laniak, G. F., Wolfe, K., & Purucker, T. (2016). Environmental models as a service: enabling interoperability through RESTful endpoints and API documentation. In *Proceedings, 8th International Congress on Environmental Modelling and Software, Toulouse, FRANCE* (Vol. 8). International Environmental Modelling and Software Society.
28. Gaie, C. (2020). From secured legacy systems to interoperable services (the careful evolution of the French tax administration to provide new possibilities while ensuring the primary tax recovering objective). *International Journal of Computational Systems Engineering, 6* (2), 76–83. https://doi.org/10.1504/IJCSYSE.2020.111204.
29. Pautasso, C., Zimmermann, O., & Leymann, F. (2008). Restful web services versus "big" web services: making the right architectural decision. In *Proceedings of the 17th international conference on World Wide Web* (pp. 805–814). Association for Computing Machinery. https://doi.org/10.1145/1367497.1367606
30. Pillai, S., Iijima, K., O'Neill, M., Santoro, J., Jain, A., & Ryan, F. (2021). Gartner Magic quadrant for full life cycle API management. https://www.gartner.com/en/documents/4006268.
31. Gaie, C. (2021). Optimise data exchange between government administrations while ensuring privacy rights. *International Journal of Business Intelligence and Systems Engineering, 1*(4), 347-357. https://doi.org/10.1504/IJBISE.2021.122786
32. Digital Government Factsheet. (2022). *France.* https://joinup.ec.europa.eu/sites/default/files/inline-files/DPA_Factsheets_2022_France_vFinal_0.pdflastvisit02/23/2023.
33. Zhang, Y. (2004). A multilayer IP security protocol for TCP performance enhancement in wireless networks. *IEEE Journal on Selected Areas in Communications, 22*(4), 767–776. https://doi.org/10.1109/JSAC.2004.825993.
34. Moura, R., Lopes, R., Matos, D. R., Pardal, M. L., & Correia, M. (2023). MultiTLS: using multiple and diverse ciphers for stronger secure channels. *Computers & Security, 132*, 103342. ISSN 0167-4048, https://doi.org/10.1016/j.cose.2023.103342.

The manufacturer's authorised representative in the EU is Springer Nature Customer Service Centre GmbH, Europaplatz 3, 69115 Heidelberg, Germany. If you have any concerns regarding our products, please contact ProductSafety@springernature.com

Printed and bound by CPI Group (UK) Ltd, Croydon, CR0 4YY
02/10/2026
02227132-0005